To the world of defect-free soldering

Measures against defects that can be made at production sites

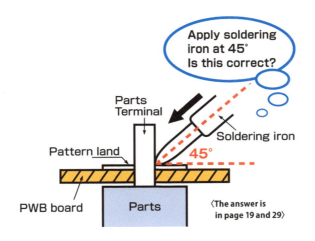

⟨The answer is in page 19 and 29⟩

Author : **Minoru Kishindo**

《 Table of contents 》

1

Preface..3

2

Soldering education (theory & practice) 5

    (2-1) History of "Soldering" (outline) 5

    (2-2) Solder material 7

    (2-3) Environmental problems of solder...................................12

    (2-4) Theory ··· Four elements of soldering.............................14

        (1) Active element: Roles of heat (temperature)...................16

        (2) Active element; Roles of the flux....................................20

    (2-5) Practical skill ··· Manual soldering (soldering iron work).....27

        (1) Roles of soldering iron (purposes)...................................27

        (2) Basic work of soldering iron..31

        (3) Method of training soldering iron work 34

        Supplement: Education for process inspectors 37

3

Manual soldering: Various know-how 40

    (3-1) Longevity of soldering iron tip 41

    (3-2) Pouring solder..43

    (3-3) Moving soldering iron more than necessary....................44

    (3-4) Scattering of "solder balls" (dispersion) 44

    (3-5) Method of repairing defective bridge between IC leads......48

    (3-6) Method of repairing soldering horn 50

    (3-7) Replacing chip parts 50

4

Measures against defective SMT reflow soldering 54

    (4-1) Chip 'Displacement', 'Floating' and 'Standing' 55

        (1) Cause of defects 55

        (2) Countermeasures 58

            1) Make the large pattern smaller (Knowhow at the

                pattern design stage) 59

            2) Review of temperature setting (Knowhow of the

                Manufacturing Division)...63

To the world of defect-free soldering

(Note 1) Is the temperature profile correctly measured?.....73
(Note 2) What happens if the chip is displaced from
the mount?..75
(4-2) 'Solder ball' defect..80
(1) 'Side ball' cause and measures.................................81
(2)'Solder ball' cause and measures................................88
(4-3) 'Less solder' cause and measures 94
(4-4) 'Bridge' cause and measures 98

5

Measures against defects of 'Flow (DIP) soldering' 102
(5-1) 'Bridge' cause and measures 103
(5-2) Cause of 'No solder' and measures................................110

6

Soldering mystery collections (Who is the criminal?).......................116
Examples of measures against defects on site
(Episode 1) Lines A and B: Why are there many customer
complaints only about boards produced in
Line A?...117
(Episode 2) Suddenly, a 'chip crack' occurs................................120
(Episode 3) Mystery of 'Print misalignment' on flexible board 125
(Episode 4) Does the characteristic of solder change
suddenly!? 131
(Episode 5) Mystery of terminal IC "No solder adheres"...........135
(Episode 6) Mystery of production plant late at night................139
(Episode 7) Suddenly, 'board swelling' happens...........................141
(Episode 8) The display of the timer disappeared suddenly
three years after buying!?......................................145
(Episode 9) Suddenly, a pattern is lost. 147
(Episode 10) Defective goods are full of treasures!
(Important know-how hidden there)..........................149
(Episode 11) Automatic driving, is it okay?................................153

The author's self-introduction..155

1　Preface

　　This book is the basic material for learning "**soldering technology**". If you master the technology (or skill) described here, your production site will become "Zero soldering defects". In addition, this book was produced with the following purposes. This book is useful for the person in the soldering production site (operator, person in charge of quality, or industrial engineer), who undertakes measures against defects and quality improvement every day.

　　It will be helpful for you for sure.

　　I wrote down about main soldering defects and the countermeasures that I experienced at the production sites in Japan and abroad for approximately 40 years. It is important to investigate and analyze the cause of the defect thoroughly. However, in many cases, wrong measures are taken without the cause of the defect being thoroughly analyzed in actual production sites. Easy action results in further defects or failures. If soldering defects occurred,do not make haste. Rather, spend time in the cause analysis. A true cause is hidden in the soldering defects. If a true cause is detected, the countermeasures are not difficult.

(A thorough cause analysis is very important)
"We understand it not because things are there but because we look at it closely."

　　With the latter, the number of defects surely becomes "0". For instance, at the quality meeting in the soldering production section, when the defect rate is 1% or 2%, we often conclude, "solder is bad" or "parts are bad". The majority of these printed wiring boards (PWBs) are supposed to become defective as long as the same "solder" is used for them if "solder" is really the cause. Don't you think so? At a defect rate of 1%, investigate a cause other than "solder". This

To the world of defect-free soldering

will lead to true measures earlier. This book is made to grasp a true cause in such a manner. For defects, representative items with a high occurrence frequency are described. (Less frequently defective items will be explained at another occasion) If you can eliminate defects by referring to this book, I am very glad.

The first half of this book is about 'Soldering education (theory & practice)'. The latter half is about 'Measures against defects' of reflow soldering and flow soldering. There is nothing difficult about 'solder' itself in this book.

In addition, I am going to produce the basic "soldering education" (DVD) for employees in reference to the first half of this book. This book can be used in combination with the video for more effective employee education. In addition, there are 'Coffee break' corners in this book. Trivia in the corners may be very useful in production sites.

(Note) The board (to which insertion electricity parts, chip parts, and IC parts, etc. are soldered) used in this book is described as PWB. PWB is an acronym of Print Wiring Board.

2 Soldering education (theory & practice)

(2-1) History of "Soldering" (outline)

Currently, solder containing lead (Pb) is hardly used, and a lot of solders without lead are used.

The history of soldering is very old. There are already signs that "Solder" was already used in the Bronze Age when the human race had begun to use metals. "Soldered" ornaments of animals were excavated from ruins of about 5,000 years ago in Persia, Egypt, and Greece. The outline history related to soldering is shown in [Fig. 2-1]. This figure shows that "Solder" has been closely linked with the human race until the present age.

In modern age, the soldering technology was 'Soldering iron work' at first. Soon after that, it became a 'Flow soldering technology' of automatic soldering devices in the age of mass production and automated production. The products have become miniaturized and densified, leading to 'Reflow soldering' that is the main current now. Cellular phones, digital cameras, and the notebook computers, etc., which are used by a lot of people, are produced by the "Reflow soldering technology" now.

Moreover, lead (Pb) that harms the natural world and the human body was included in 'Solder' materials for a long time. A lot of people all over the world involved in solder had continued efforts, leading to the first-ever lead-free solder material (lead-free solder) in around 2,000 AD. Currently, solder containing lead (Pb) is hardly used, and a lot of solders without lead are used in production. Lead-free solder is a wonderful material that does not harm the natural world and the human body.

To the world of defect-free soldering

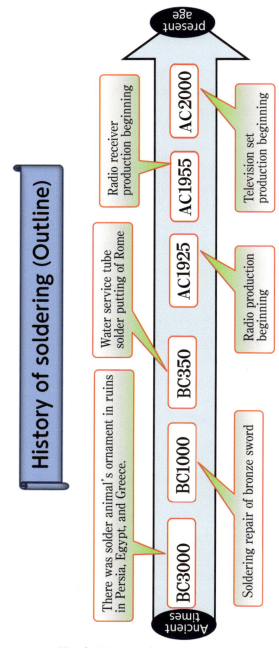

[Fig. 2-1] History of soldering (Outline)

(2-2) Solder material

'Soldering' is to connect two different metals mechanically and electrically. The human race has used this technology for about as many as 5,000 years in various respects. Still a lot of electronic products (for instance, mobile phones, digital cameras and notebook computers, etc.) have become final products by soldering. And we might not have been able to obtain convenient modern tools in the civilized life if the material called 'Solder' had not existed in this world. The materials of solder had been an alloy of 'tin (Sn)' and 'lead (Pb)' for a long time. However, lead (Pb) negatively affects the earth environment and human body. Therefore, a lot of lead-free solders have come to be used. In general, the differences between lead-free solders and Sn-Pb solders are as follows.

	SOLDER	Composition	Melting point
1	Lead-free solder	Sn62% , Pb38%	183℃
2	Sn-Pb solder	Sn-3Ag-0.5Cu	217℃

Please pay attention to the difference of the melting points. The state chart about 'Sn-Pb solder' is shown in [Fig. 2-2]. The melting point of Sn is 232℃, and that of Pb is 327℃. The melting point of the Sn62/Pb38 alloy (note 1) falls down to 183℃ (eutectic point), and we have used this alloy as 'Solder'. This solder is called 'Eutectic solder'.

(Note 1) The Sn62/Pb38 alloy is composed of Sn61.9% and Pb38.1% in more detail.

The state of the Sn-Pb alloy is simple and basic though 'Sn-Pb solder' will disappear in the near future. Understanding this state chart [Fig. 2-2] will be very useful for soldering in the future.

The solder material that the human race has used for a long time, as shown in [Fig. 2-2], has been the solder alloy of point b. (point b is a eutectic point.) Pay attention to point b (eutectic point). It is a liquid state at 183 ℃ or higher and a solid state at less than 183 ℃.

To the world of defect-free soldering

Point a is 327°C in melting point at Pb100%, and point c is 232°C in melting point at Sn100%.

The melting point falls at point b, making it easier to handle solder materials.

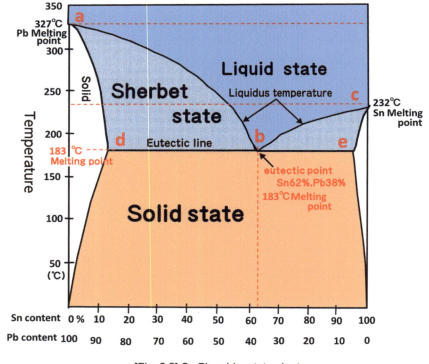

[Fig. 2-2] Sn-Pb solder state chart

'Soldering' is to connect two different metals mechanically and electrically. Base metals other than gold (Au) are oxidized (initial state of rust) in air. Then, they become rusted.

For instance, when copper (Cu) is left in air, the surface becomes CuO. CuO is an oxide film and an initial state of rust.

2 Soldering education (theory & practice)

This oxide film (CuO), hardly visible with the naked eye, is always generated on the surface of the metal.

The mechanism to solder the surface of copper (Cu) covered with the oxide film is as shown in [Fig. 2-3].

When this oxide film (CuO) exists, it is absolutely impossible to solder. However, the material called flux is always used in the soldering work, removing the oxide film chemically and completely.

(The working of flux will be explained in detail in the following chapter. It is an attractive and magical material.)

Sn and Pb atoms in solder move freely with the heat (about 300°C) generated by soldering. The oxide film (CuO) in right figure B becomes liquid because of the action of the flux, and will disappear with time. (C)

The Sn atoms of the solder and the Cu atoms of the base metal move and collide mutually. Moreover, the Sn atoms in the solder enter into the base metal Cu. (D) As a result of D, intermetallic compounds Cu_3Sn (alloy layer) and Cu_6Sn_5 (alloy layer) are generated. (E) Soldering is completed by the occurrence of phenomena from A to E.

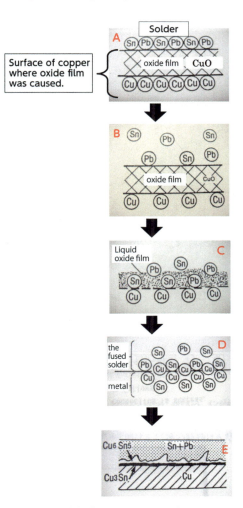

[Fig. 2-3] Soldering mechanism

To the world of defect-free soldering

Easy or difficult to solder: About various base metals

There are a metal that solders easily and a difficult metal. [Fig. 2-4] shows the levels. As shown in the figure, tin is the easiest to solder.

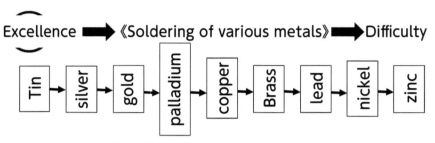

[Fig. 2-4] Rank of metal that is easy to solder

Usually, plate the surface of the metal difficult to solder for easier soldering. There are the following kinds of plating, used for surface treatment of a lot of metals.

1. Solder plating
2. Tin plating
3. Silver plating
4. Gold plating

Respectively, the shape of the solder material is different depending on the production site used.

For instance, creamy solder used in the SMT production site is put in the container.

Please refer to [Fig. 2-5] for a solder material with a different shape.

2 Soldering education (theory & practice)

《 Externals of solders 》

	Type of solder	Remarks
1	Flux cored wire solder 	It is used for soldering iron work. (Manual Soldering)
2	Bar solder 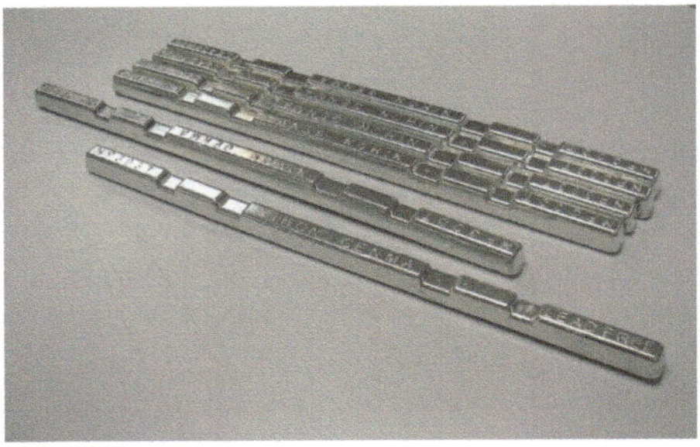	It is used at a flow solder bath. (Flow soldering)
3	Cream solder 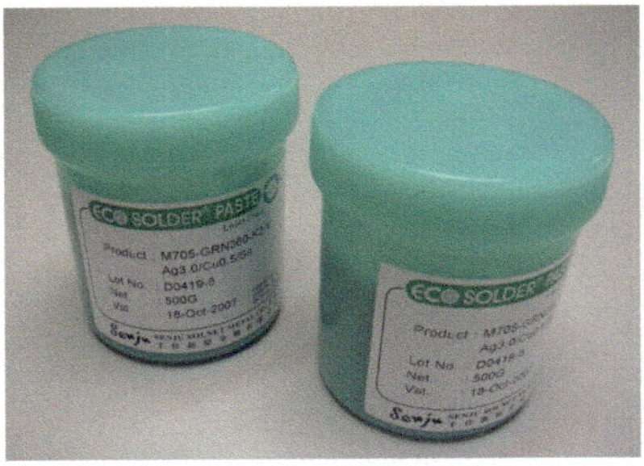	It is used at a reflow furnace. (SMT reflow soldering) SMT:Surface Mount Technology

[Fig. 2-5] Kind of solder material

11

To the world of defect-free soldering

(2-3) Environmental problems of solder

The human race has used "Solder" for about 5,000 years in various places, and people have become more affluent as generation progressed. However, "lead (Pb)" included in solder was found to be harmful to human body and the environment, and the restriction and legislation on the solder material started in recent years (from year 2000). A PWB (note: page13) using solder with lead as shown in [Fig. 2-6] is disposed of, the soil is polluted with lead by rain water. In addition, lead infiltrates underground water and finally, even drinking water is polluted with lead, and lead enters and is accumulated in human body. When a large amount of lead is absorbed into the human body, it damages a brain and nerves, and a child's growth is inhibited.

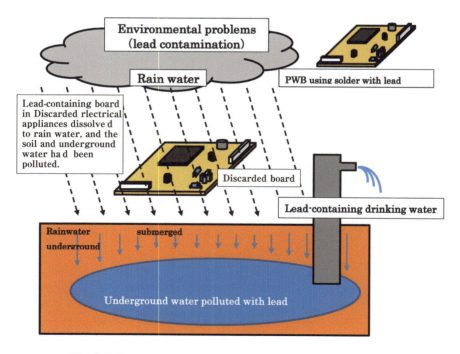

[Fig. 2-6] The environmental pollution of solder that contains lead.

2 Soldering education (theory & practice)

Thus, it became more and more difficult to use solder that contained lead for fear of environmental pollution. Use of solder that contained lead was restricted to prevent such environmental pollution, and solder manufacturers have continued the research of solder that does not contain lead for years. As a result, solder without lead (Pd free solder) has been used for boards of various products since around year 2000.

(note) PWB stands for Printed Wiring Board.

To the world of defect-free soldering

(2-4) Theory ⋯ Four elements of soldering

I do not intend to say anything difficult about solder. However, the knowledge of the '**four elements of soldering**' is absolutely necessary for persons in the production site for soldering. Please do not forget it.

Four elements of soldering
A: Heat (Temperature)
B: Flux
C: Metal
D: Solder

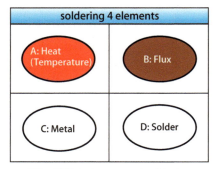

[Fig. 2-7] 4 elements of soldering

It is not possible to solder if even one of the four elements is missing. (Refer to [Fig. 2-7])[Fig. 2-8] is an expanded cross section of the flux cored wire solder. The central white part is the flux, and there is the flux in the central part wherever you cut. For example, soldering without B flux was tested. The result was as shown in [Fig. 2-9] and [Fig. 2-10] on page 15. The right of the photograph is without flux. In [Fig. 2-9], on the left side of the photograph the wire solder was cut short and placed on the pad. (tested with flux) The wire solder whose surface was cut was placed on the pad on the right side. (tested without flux) Both solders were placed on the same board and passed through the reflow furnace. The result was as shown in the lower photograph in [Fig. 2-9]. The left side is soldered on the pad without trouble. The right side is not soldered, and has come out from the reflow furnace like a dog's excrement. Likewise, test results with and without flux in the cream solder are as shown in [Fig. 2-10]. The test result without flux shows the state like solder ball particles. Thus, it is not possible to solder if flux, one of the four elements, is missing. This flux will be explained in detail later in 'Role of the flux'

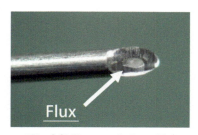

[Fig. 2-8] Flux cored wire solder

2 Soldering education (theory & practice)

Soldering test with flux and without flux

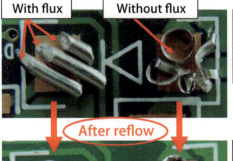

- With flux
- Without flux
- Small pieces of the wire solder on the pad
- The surface of wire solder is cut down with the cutter

After reflow

- It is soldered.
- It is not soldered.

[Fig. 2-9] Test of wire solder

- The cream solder is placed on the pad.
- Cream solder without flux

After reflow

- It is soldered on the pad.
- The solder particles on the pad.

[Fig. 2-10] Test of cream solder

15

To the world of defect-free soldering

Each of the four elements is not equal with one another. In fact, the four elements are divided into two categories. (Refer to [Fig. 2-11]) One consists of active elements, and the other consists of passive elements. As shown in [Fig. 2-11], the active elements are A heat (temperature) and B flux, and the passive elements are C metal and D solder.

"Metal" and "Solder" are independent and it is impossible to solder only by themselves. These are passive. Therefore, they become passive elements. However, "Heat (temperature)" and "Flux" influence "Metal" and "Solder" in various manners, and soldering is completed. Therefore, these become active elements.

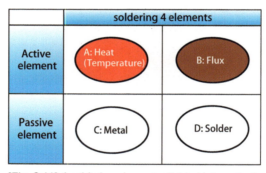

[Fig. 2-11] 4 soldering elements divided into actively and passive

(1) Active element: Roles of heat (temperature)

In soldering, heat (temperature) plays very important roles as an active element. Why is it "Active"? Because of the importance of supplying heat (temperature) to the metal to be soldered. However, the temperature is apt to be overlooked from the items of causes of troubles because it is not visible immediately on the production site. Therefore, "parts" or " PWB" was often referred to as a cause of a trouble. As a result of a thorough investigation of a defective item, the cause was often related to temperature like an insufficient temperature or a problem in the measurement of the temperature. Soldering is not completed if temperature (240-250℃) is not added

2 Soldering education (theory & practice)

to the metal to be soldered. The left of [Fig.2-12] shows soldering without problem. But, the right resulted in defective soldering due to an insufficient temperature of the terminal (240℃ or less).

Soldering is to tie two metals mechanically and electrically, but the two metals are not the same in size (not same heat capacity). **Difference in size of two metals. → Difference in temperature rise → Defective soldering on a metal where the temperature is low.** The temperature of the pattern pad (small metal) rises more quickly than the part terminal (large metal) in [Fig. 2-12]. It is important to apply the temperature of about 240-250℃ to a large metal quickly so as to make an intermetallic compound (hereinafter, alloy layer). (Note) See page 24 《 Completion of soldering 》 about the alloy layer. A defect is often caused by **insufficient heating of a metal having large thermal capacity.**

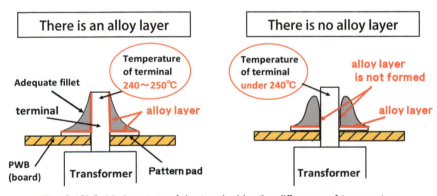

[Fig. 2-12] Soldering state of the terminal by the difference of temperature

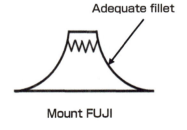

Please pay attention to action of a. A heat (temperature), in [Fig.2-13]. A actively applies heat to a metal to be soldered and raises its temperature. If the metal temperature is raised to about 240-250℃, the alloy layer can be formed

To the world of defect-free soldering

between the metal and the solder, and an adequate solder fillet is formed. If the temperature applied to the metal is insufficient, an alloy layer is not formed in the metal, resulting in a soldering defect. (See the presence or absence of the alloy layer in [Fig. 2-12].)

Soldering is to form the alloy layer (**Cu$_3$Sn** in the case copper) to the metal with any of manual soldering, flow soldering and reflow soldering. Usually, the temperature of a soldering iron tip is set to 350-360°C. Wire solder melts immediately at this temperature (solder melting point: 217°C). Therefore, pay more attention to **a** raising the temperature of the metal rather than melting the wire solder. About **b** in [Fig. 2-13], the melting point of the solder (Pb-free solder) is 217°C. Therefore, the solder melts without trouble because soldering iron tip temperature, flow soldering temperature and reflow soldering temperature are set at 217°C or more. **c**, the "activation of flux" is to activate the flux within the range of the temperature (preheating range) below the melting point of the solder. The action of the flux will be explained in detail later in "Role of the flux".

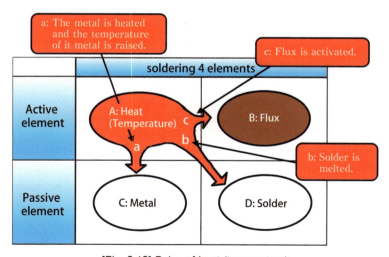

[Fig. 2-13] Roles of heat (temperature)

18

2 Soldering education (theory & practice)

Zero defect using the action of a in the production site!

The cover of this book shows how to apply the tip of the soldering iron. The temperature of the small pad of the PWB rises quickly. The temperature of the big metal (for example, a part terminal) does not go up as quickly as the PWB pad. When the angle of the soldering iron is 45 degrees, the temperature of the part terminal rises more slowly than that of the board pad. As a result, the temperature of the terminal is low, an alloy layer is not formed, resulting in a soldering defect. To avoid the defect, it is necessary to apply the soldering iron to the terminal at the angle of about 60° to 70° as shown in [Fig. 2-14]. Then, the temperature of the terminal goes up quickly to 240-250℃. And an alloy layer is formed. Moreover, the worker has to instantly judge the angle of this soldering iron as soon as checking the soldering point.

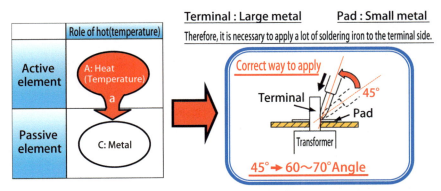

[Fig. 2-14] Role of heat(temperature) of soldering iron and correct way to apply soldering iron.

In the soldering work, it is often said, "Make a fillet of a brazed joint (skirt shape of Mt. Fuji)". A solder fillet is formed as a result of applying heat (temperature of 240-250℃) to the metal to be soldered. On a metal to which heat is not applied (temperature is not raised), a fillet is not formed by any means.

To the world of defect-free soldering

(2) Active element; Roles of the flux.

The flux has two big roles, which are **d** and **e** of [Fig. 2-15].

d: An oxide film is chemically removed from the surface of the metal and that of the solder.

e: The surface tension of the molten solder is reduced, thereby promoting wetting of the solder.

I have been engaged in the production site of soldering long and understood well how wonderful the material of flux is. It is not an exaggeration to say, "**flux is a magic material**". Use the magic material (flux) wisely. Comducted the above-mentioned soldering tests with the flux and without the flux. (See page 15) As a result of the tests, it was not possible to solder without flux. The reason is that an oxide film on the metal surface was not successfully removed. Pay attention to the action of d in [Fig. 2-15]. All metals in the air are oxidized (except gold). A metal under high-temperature status in a reflow furnace is intensely oxidized. Flux removes the oxide film chemically, finely,and magically.

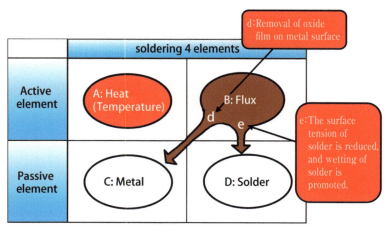

[Fig. 2-15] Roles of the flux

2 Soldering education (theory & practice)

However, three roles of flux are often mentioned. Besides the above-mentioned **d** and **e**, there is another role, "Flux covers the surface of the metal and prevents its re-oxidation while soldering." See [Fig. 2-16] for this role. When the metal and the solder are surely covered with flux, this metal is not re-oxidized. However, the soldered part is found to be not fully covered with flux if carefully checked with a magnifying glass. In this case, the soldered part is re-oxidized. See [Fig. 2-16]. For example, when the low-temperature solder paste containing Bi (bismuth) is used, there sometimes occurs a "micro bridge" (a super-thin solder bridge which is hard to check with the naked eye). This became defective due to the re-oxidation. Therefore, it seems that "Flux prevents re-oxidation" is wrong.

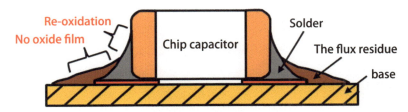

[Fig. 2-16] After reflow, with oxide film, nothing

After the reflow, there are the part covered with flux and the part not covered in the cream solder that hardens as shown in [Fig. 2-16]. The part not covered with flux is re-oxidized while the part covered is not oxidized.

To the world of defect-free soldering

d: Elimination of oxide on metal surface.

All metals other than gold (Au) are surely oxidized in the air. For example, in PWB land (Cu), if there is the oxide film (CuO), soldering is impossible by 100%. Heat to remove this oxide film will require the temperature of 1,000℃ or more. All electric parts on the PWB are damaged if the temperature reaches 1,000℃ or more. As a countermeasure at that time "Magical substance, flux" is the emergence! Flux removes an oxide film (CuO) chemically and finely. In the above-mentioned, "Soldering tests with flux and without flux" [Fig. 2-9, 10], the one that was not soldered came out of the reflow oven in the state of i) with the oxide as shown in [Fig. 2-17]. On the contrary, the soldered one changed as shown in i) → ii) → iii) → iv) in the furnace,

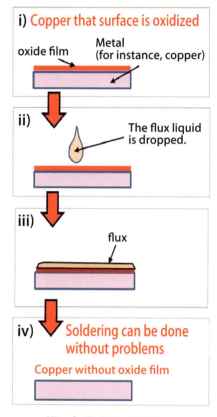

[Fig. 2-17] Role of the flux

and the oxide film was removed. As a result, the quality item came out of the furnace by the soldered state. Therefore, **the role of flux is to remove an oxide film on the metal surface.**

e: The flux lowers the surface tension of the molten solder and promotes wetting of the solder.

The following are explanations of **e** in [Fig. 2-15]. **The surface tension** of a liquid is force to become a globular form. Specifically, (refer to Fig. 2-18) it is the force to become round like a table tennis ball in the liquid state. When molten solder is compared with water,

22

2 Soldering education (theory & practice)

of course both of them are liquid. They have surface tension each. Because surface tension of the molten solder is larger, it is rounder than water as shown in [Fig. 2-18]. "The surface tension of the molten solder becomes small" means that the molten solder on the left side of [Fig. 2-18] becomes close to the state of right-hand side water. See [Fig. 2-19]. i) If flux liquid is applied to the molten solder, the globular shape collapses as shown in ii), and further it collapses as shown in iii), and it becomes closer to the shape of water. As mentioned above, small surface tension promotes "wetting" of the solder and is important phenomenon for soldering.

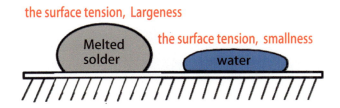

[Fig. 2-18] Large and small in surface tension

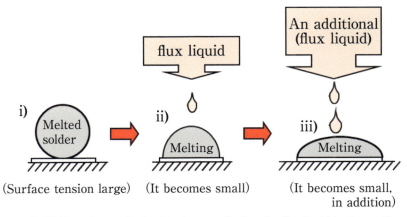

[Fig. 2-19] How does melted solder become it when the flux liquid is dropped?

"Flux, a magic material", as mentioned above, removes an oxide film and reduces surface tension. Flux can work in these different functions without trouble. It is exactly "The magic material". A person

To the world of defect-free soldering

who controls flux will control soldering. Let's use flux as much as possible!

《 Completion of soldering 》

Refer to [Fig. 2-20]. If active elements of A (Temperature) and B (Flux) fully play a role, an alloy layer will be formed between C (Metal) and D (Solder), which are passive elements. Solder is wet and spreads, and **soldering is completed**.

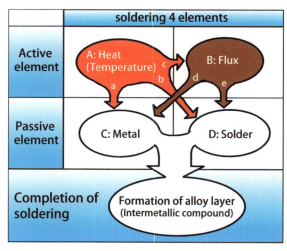

To "Complete soldering" without the problem in the quality, it is only necessary that active elements (a,b,c,d,e) of A and B work sufficiently.

[Fig. 2-20] Completion of soldering

As above, the explanation of the theory of the 'four elements of soldering' is completed. In order to "solder without defect", it is important to surely master the roles of these "4 elements of soldering" and employ the roles efficiently and practically. From the next chapter, actual soldering, soldering iron work, flow soldering and reflow soldering will be explained.

2 Soldering education (theory & practice)

Relax and coffee time (1)  **The problem of flux.**

Although only the merit of flux as the **"magic material"** has been described, there is a demerit, too. It is a problem of a flux residue on a PWB whose soldering was completed. Long before, the soldered PWB was washed, as common sense, by Freon liquid, and the flux residue was washed out. Now, due to the environment problems, it has become impossible to use Freon gas and soldered PWBs have not been washed. However, if the flux residue is left on the board for a long time, it absorbs moisture in air, causing a chemical reaction of flux. (Chemical reaction: Generation of migration and whisker) As a result, a short circuit occurs between adjacent patterns, causing a big problem. Such problems have been reduced as flux has been changed from an old **RA** type to the current **RMA** type thanks to solder makers' efforts.

Flux of **RA** type ⋯ Strong activity flux
Flux of **RMA** type ⋯ Weak activity flux

These problems, even if reduced, seem to exist in a high density board.

To the world of defect-free soldering

《Soldering iron, name of each part》

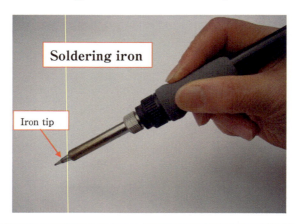

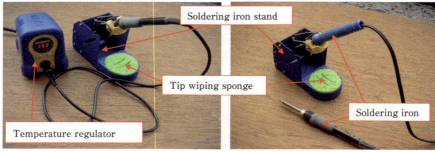

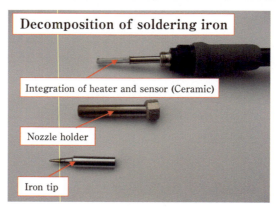

(From HAKKO's products)

(2-5) Practical skill ··· Manual soldering (soldering iron work)

Soldering iron work is also known as "**manual soldering**". Because manual soldering is not automatic, it is called "manual". Because mass production is carried out in the production plant, automatic soldering processes, such as SMT reflow soldering and flow soldering, are the mainstream, and there are fewer opportunities to use soldering iron work (manual soldering). However, manual soldering is still the mainstream for adjustment and repair of defective goods in the production process. Unlike automatic soldering equipment, such as a reflow oven and a solder bath, quality of soldering iron work will be greatly affected by the worker's skill level. Therefore, most factories position manual soldering as **a special process** and establish a "**qualification system**" for education. It is very important to continue such in-house education and training to improve workers' skill levels in terms of eliminating complaints from customers and the market.

An educational video of the soldering iron work will be produced shortly. Please use this book and the video for in-house training.

(1) Roles of soldering iron (purposes)

First of all, you are requested to change your perception about the "Roles of soldering iron (purposes)".

Usually, the temperature of a solder iron tip is set at about 330℃. The melting point of flux cored wire solder is about 220℃. Therefore, if the wire solder is applied to the iron tip, naturally it will melt. However, the temperature of the iron tip has another important purpose.

Watch carefully how workers do their job in the production area. They apply the iron tip to it immediately to melt the wire solder. This is a big mistake.

> 1 The soldering iron is not a tool to melt solder.

The workers who regard the soldering iron as "the tool to melt the solder" cause defects during the work.

It is not a tool to melt solder.

To the world of defect-free soldering

| 2 The soldering iron is a tool to heat the metal to be soldered. |

This metal is to be soldered from now on. It is very important to do the soldering work every day after having understood "**2**". When you do the soldering iron work, always bear "**2**" in mind. If the metal temperature is raised to about 240-250℃ and wire solder is applied to the junction, an alloy layer is always made. Then, it is possible to solder with sufficient quality. Even if a wire solder is melted only with the soldering iron without the metal temperature being raised, an alloy layer is never formed in the junction by any means. The cause of defective soldering called "Cold solder" is definitely due to the work of "**1**" being done unconsciously.

《 How to apply the soldering iron correctly? 》

4 procedures shown in [Fig. 2-24] have been explained in the training about how to do the soldering iron work. A defect possibly occurs after the operation only with these four procedures. There was a question as shown in [Fig. 2-21] on the cover of this book. (The angle of the soldering iron is 45 degrees) Although the answer was described also in page 19 of "Theory", further details are given here. Normally, soldering work has to be completed in less than 4-5 seconds on the production site. The soldering iron is applied at the angle of 45 degrees in such a short time. Is this really good? Sizes of two metals soldered are different. (Examples here are a part terminal and a pattern pad)

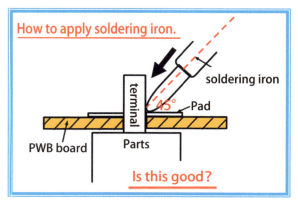

[Fig. 2-21] How to apply soldering iron.

2 Soldering education (theory & practice)

Therefore, the two metals have different quantities of heat. The temperature of the larger metal rises slower than that of the smaller metal. Therefore, an angle of the soldering iron needs to be changed for the larger metal so that the temperature rises early. When a soldering iron worker sees the metals of the terminal and the pad, he/she needs to instantaneously judge which metal is larger. It is a big problem to apply an iron tip unconsciously. (You have to instantaneously judge the larger and smaller thermal capacities) When the foot (terminal) of the transformer is judged to be a metal bigger than a pad in the case of [Fig. 2-21], the iron tip has to be applied to the terminal side in the angle of 60-70 degrees instead of the angle of 45 degrees as shown in [Fig. 2-22] so as to raise the terminal temperature to 240-250℃ early for forming an alloy layer. Moreover, the alloy layer has to be formed within 5 seconds of operation time.

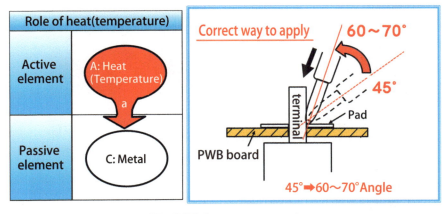

[Fig. 2-22] Correct way to apply

To the world of defect-free soldering

[Fig. 2-23] is a comparative drawing of a method of applying the soldering iron and its result.

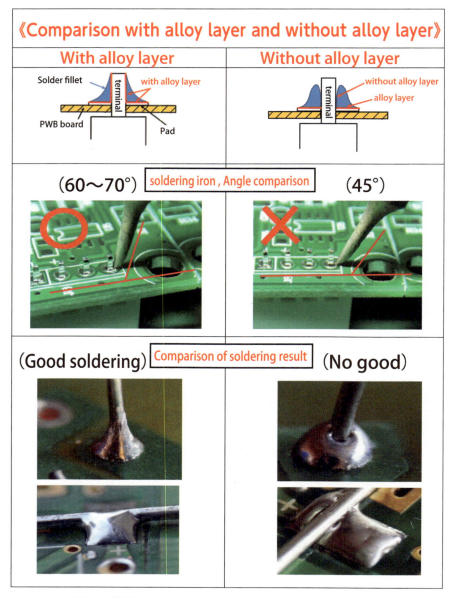

[Fig. 2-23] Comparison with alloy layer and without alloy layer

2 Soldering education (theory & practice)

(2) Basic work of soldering iron

In addition, the next is the practice of soldering iron work. It is requested to practice the following four procedures as the basic for practicing the soldering iron work. Even in view of the instructions and the roles of the soldering iron, it is insufficient only to instruct the following **four procedures** for preventing a defect.

1. <u>Apply the soldering iron to the portion to be soldered.</u>

2. <u>Add wire solder.</u>

3. <u>When proper quantity of solder is added, remove the wire solder.</u>

4. <u>Remove the soldering iron.</u>

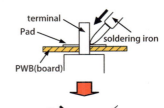

[Fig. 2-24] Four steps of soldering iron work

Six procedures shown in page 32 are necessary to do correctly to maintain quality from the time of daily training. The ideas of "Roles of soldering iron" have been introduced in **the six procedures**. Therefore, it is necessary to thoroughly understand the six procedures to repeat practice of the soldering iron work. It is important to engage in the soldering iron work is important after understanding the six procedures very well and practicing them.

To the world of defect-free soldering

《6 soldering iron works don't put out defects》	《4 soldering iron works put out defects》
1. Judge the largeness and smallness of two metals instantly. 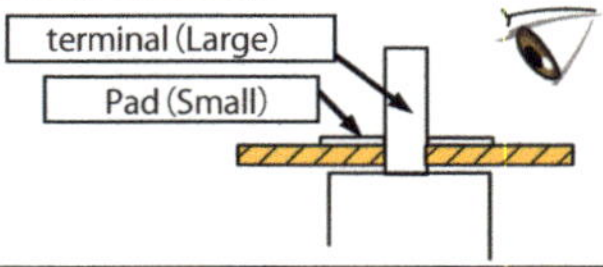 terminal (Large) Pad (Small)	
2. Apply the soldering iron to the portion to be soldered.  Apply a lot of soldering iron to the terminal side, and raise the temperature of the terminal.	1. Apply the soldering iron to the portion to be soldered. 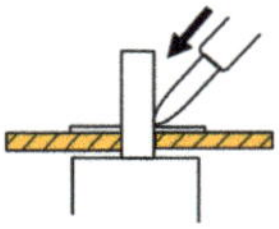 It is problem work that applies soldering iron without considering the largeness and smallness of two metals.
3. See time where the temperature of the terminal goes up, and add the wire solder. 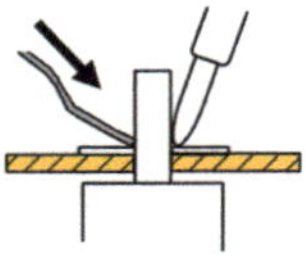	2. See time where the temperature of the terminal goes up, and add the wire solder. 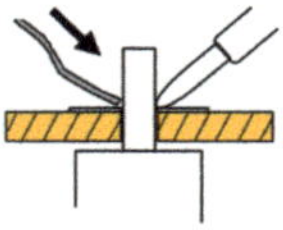
4. Remove the wire solder when you add the proper quantity solder. 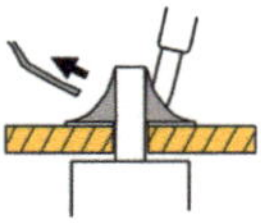	3. Remove the wire solder when you add the proper quantity solder. 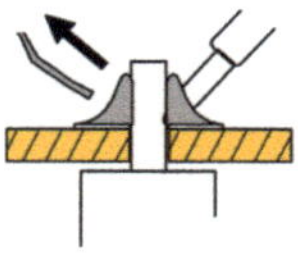
5. Remove the soldering iron. 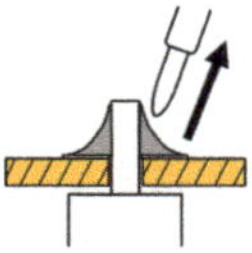	4. Remove the soldering iron. 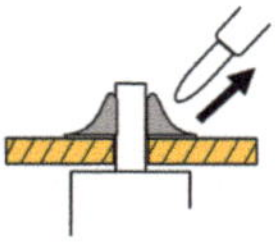
6. Visual inspection after the working. (Confirmation) 	

[Fig. 2-25] Correct soldering iron work and problematic soldering iron work.

2 Soldering education (theory & practice)

Additional procedures **(1 and 6)**

    1 ⋯ Immediately judge the larger from two metals, and add temperature to the larger one.

    6 ⋯ Visually inspect an "operational error" after completing soldering work

《 Soldering work has to be completed while smoke appears 》

When flux cored wire solder is melted with soldering iron, smoke is generated. (Refer to [Fig. 2-26]) What is this smoke? This is smoke that rises when the flux in wire soldering is burnt at the temperature of soldering iron (about 350℃). Where is the flux in the fine wire soldering? Refer to [Fig.2-26]. The flux is the whitish one in the center of the section diagonally cut. Smoke doesn't rise if it takes ten seconds or more for soldering iron work. The flux becomes smoke and disappears. Soldering without flux is not lustrous and becomes fragile. (One of the four soldering elements disappeared.) Therefore, it is necessary to end soldering iron work within 4-5 seconds while smoke is rising.

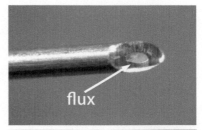

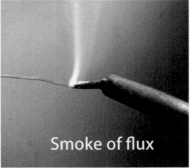

[Fig. 2-26] Flux and smoke

To the world of defect-free soldering

(3) Method of training soldering iron work

The good method of concrete training is to make "**Net lines**" as shown in [Fig. 2-27]. This training is good for practicing as reference. Solder the part at the intersection showing a cross of lines sequentially. Apply the soldering iron to both metals in the cross. And if solder is added after raising the temperature of both metals, a beautiful

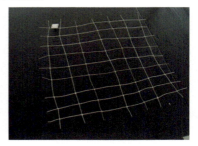

[Fig. 2-27] Net line making

solder fillet will be formed. The worker can master the skill to apply soldering iron to both metals to raise the temperatures. In addition, the worker uses the wire rods having different thicknesses (0.8φ and 1.2φ) to solder the net line. At that time, the worker can experience which wire is closer to the soldering iron tip. This is a valuable experience for them. After completing this practice, learn how to detach and install chip components as part of soldering iron work.

《 Preparation to make net lines 》

1. The solder plating lines (0.8φ, 1.2φ) that are being marketed. [Fig. 2-28]
2. Cut the solder plating lines of 1 in the length of 30 cm (18 x 0.8φ lines). [Fig. 2-29]
3. Cut the solder plating lines of 1 in the length of 30 cm (0.8φ x 12 lines and 1.2φ x 6lines). [Fig. 2-29]

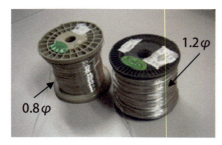

[Fig. 2-28] Plating Wire

[Fig. 2-29] Wire rod of 30cm

2 Soldering education (theory & practice)

First of all, it is recommended to start from **2** for instructing the soldering iron work for a beginner. There is no problem for an experienced person to start from **3**. Skill practice of **3** is soldering of wire rods with different thicknesses. And this practice is to learn how to solder without problem. At first, solder four places to become a square. (Refer to [Fig. 2-31]) At this time, it is very difficult for a beginner to handle wire rods that move. It becomes to solder relatively easily after fixing the wire rods with a "weight". Hereafter, solder the vertical line and the horizontal line in the central part as shown in the

[Fig. 2-30] Flux cored wire solder Made from Senju Metal Industry Co., Ltd.

[Fig. 2-31] Four outer frame soldering

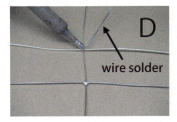

[Fig. 2-34] Making a net line

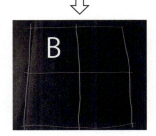

[Fig. 2-32] Middle vertical and horizontal soldering

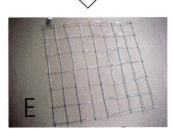

[Fig. 2-35] A net line

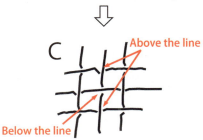

[Fig. 2-33] Method of adding wire rod

35

To the world of defect-free soldering

photograph. When you add the wire rods vertically and horizontally, add them alternately to be "up", "down", "up", "down" on the intersection as shown in [Fig. 2-33]. Finally, make net lines as shown in E.

《 Judging standards of net line work and product 》

Judging standards of practicing use of net lines:

1. Is a solder fillet formed in a joined area? (Was the temperature added enough?)
2. Does the solder have gloss? (Was work completed in a short time while flux existed?)
3. Is the solder quantity proper?(It is not good to be excessive or insufficient)
4. Doesn't solder adhere to an extra part?
   (In the four items, 1 and 2 are important in terms of the judging standards)

With the above-mentioned standards being set, a worker's skill level can be judged.

[Fig. 2-36] shows that the 0.8φ wire rod was soldered with the 1.2φ wire rod, and a beautiful fillet of junction is formed. This is a result of adding the temperature correctly.

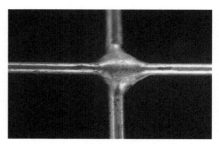

[Fig. 2-36] There is a fillet, Work OK

Look at the photograph in [Fig. 2-37]. A fillet of junction is not sufficiently formed. The wire rods were not sufficiently heated. The reason why luster is not seen is because, as a result of spending too much time in soldering, it already had no flux. In such work, there is no soldering strength.

[Fig. 2-37] Soldering with problem

2 Soldering education (theory & practice)

[Fig. 2-38] shows a product of an operator who actually made the net lines.

[Fig. 2-38] Soldering iron worker's net line sample

Supplement: Education for process inspectors

Inspectors are instructed at each factory. An effective method of education is introduced here..
1. Make internal inspection standards and educate according to them.
2. Take macro photography of defective items generated on the production site (photomicrograph, if possible). At that time, take photographs of good quality items, too. And record these photographs in a card. (Refer to following [Fig. 2-39]).
3. Clarify the standards of quality items and defective ones with the photograph card according to 2.
4. Show the photograph card in inspection personnel's training site. Ask them, "Why did you judge it as good?" and "Why did you judge it as defective?"
5. Educate the inspection personnel so as to correctly answer the question.

The above-mentioned education for inspectors is immediately reflected in inspection results. Especially, it is even more effective to use photographs in the present production site rather than generally-used ones.

To the world of defect-free soldering

Photograph cards: Good items / defective items

[Fig. 2-39] Card in photograph

2 Soldering education (theory & practice)

Relax and coffee time (2) **What a surprise!**

Sometimes I encountered a surprise on the production site. Photograph on the right: From the back of the worker. He was replacing chip parts with a soldering iron. A metallic parts plate was near the worker and contained about 20 chip

Surprising work

capacitors. What a surprise! A chip capacitor was immediately applied to the soldering iron of over 350℃, and the part was moved. In this case, grip and move the part with tweezers to the soldering place on the board rather than apply the soldering iron to the part. Apply the soldering iron to a part only during soldering. Chip parts (Ceramic material) are very vulnerable to heat. Therefore, it is necessary to avoid thermal shock on them as much as possible. When the chip parts are applied to the soldering iron of 350℃ for a long time, a crack is generated by thermal shock, resultantly causing damage on them. Such a problem cannot be found without the production site being patrolled. It is necessary to instruct the method of installing chip parts over and over again. And it is necessary to post the comparison between defective work and correct work (With a drawing or photograph) in the production site in the visible manner.

A big problem occurs in the place where leader's eyes do not reach.

To the world of defect-free soldering

| 3 | Manual soldering: Various know-how |

(3-1) Longevity of soldering iron tip

(3-2) Pouring solder

(3-3) Moving soldering iron more than necessary

(3-4) Scattering of "solder balls" (dispersion)

(3-5) Method of repairing defective bridge between IC leads

(3-6) Method of repairing soldering horn

(3-7) Replacing tip parts

The result of soldering iron work is the different depending on the skill of the worker unlike automatic soldering, such as flow or reflow soldering. Insufficient skills lead to defective goods, confusing a production site. To avoid such a problem, **the knowhow of soldering iron work** is explained. First of all, the manager and the leader on the production site have to decide the following 1 and 2.

1. **Decide an appropriate type (shape of the tip) and wattage of a soldering iron to be used.**

    As mentioned above, the soldering work on the production site has to be completed in about 4-5 seconds or less. As the role of a soldering iron is to add temperature (heat) to the metal, it is necessary to keep appropriate the wattage of the soldering iron and the shape of the tip. A soldering iron whose heat conduction is insufficient is difficult to work on, causing a defect.

2. **Decide a wire diameter of flux-cored wire solder.**

    If a wire diameter is large, temperature of the iron tip falls momentarily, and soldering work of 4~5 seconds or less also causes a defect. It is also important to use flux-cored wire solder having an appropriate wire diameter.

3 Manual soldering: Various know-how

(3-1) Longevity of soldering iron tip

A tip of a soldering iron is plated. If the hole opens or the plating peels off, its lifetime already is end and can't be used. Even if you use such an iron tip, it is still possible to solder. However, heat conduction of the iron tip is insufficient and it takes longer to solder. It becomes difficult to work, and the tip of the soldering iron moves. As a result, a pattern peels off, and flux and solder balls disperse. It is necessary to replace the iron tip which the lifetime already is end with a new one. How to judge a lifespan of an iron tip: As shown in [Fig. 3-1], when you apply the wire solder to the tip, it is possible to use it if the wire solder adheres to the tip of the soldering iron. However, the solder can melt with the soldering iron tip of ending longevity but the melting solder doesn't adhere to the iron tip. It is repelled and becomes spherical at the tip of the thread solder. (Refer to [Fig. 3-1])

[Fig. 3-1] Judgment of longevity of iron tip

To the world of defect-free soldering

Relax and coffee time (3)

While the smoke of solder is rising ⋯

Cigarette smoke can be romantic, but solder smoke is never romantic at all, showing the situation is urgent. The solder smoke is scorch of the flux that exists in the center of the wire solder. Why is it urgent? Because this smoke disappears at once and the flux finally becomes nothing. While soldering, the flux, one of the four elements, disappears, causing a big problem. In the soldering work without smoke (flux), the solder is not lustrous and becomes fragile, resultantly causing a big problem in terms of quality.

Let's finish the soldering work while smoke is rising!

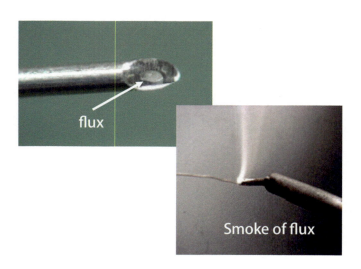

3 Manual soldering: Various know-how

(3-2) Pouring solder

[Fig. 3-2] Work on the right side is sometimes seen as problem work of a beginner on the site. The wire solder is melted with the soldering iron without adding temperature to a part terminal and a pattern pad of the soldered part. And the melted solder is poured on the pad. The temperature of the iron tip is about 350℃. Therefore, solder melts instantly and, if solder is poured, it is misunderstood that the soldering work is completed. This result is a state of "cold solder" without the formation of an alloy layer. First of all, apply the tip of the soldering iron to the metal and concentrate on adding heat (temperature). Afterward, it is important to add wire solder to the soldering part. It is usual for a beginner to do this problematic work. As explained in pages 27 and 28, **"the soldering iron is a tool not to melt solder but to heat a metal"**. It is important to understand this for soldering.

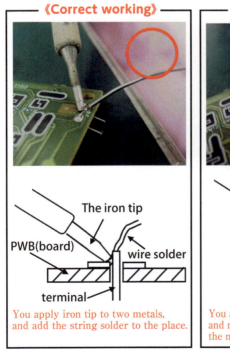
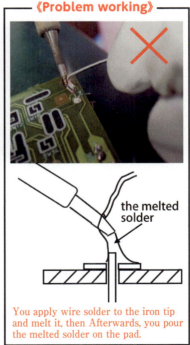

[Fig. 3-2] How to treat and problem work of soldering iron

To the world of defect-free soldering

(3-3) Moving soldering iron more than necessary

A worker with a soldering iron on the production site moves the iron up and down or right and left. The purpose of the soldering iron is to use as **"a tool to add temperature to the metal"**, so it is not necessary to move it. When it is moved, the temperature of the metal doesn't rise, causing defects and problems. Therefore, apply the soldering iron to the metal, and don't move the tip of the iron. It is necessary to raise the temperature of the metal. Moreover, avoid moving the soldering iron as much as possible so as not to disperse 'solder balls', which will be explained next. I have told not to move the soldering iron while working up to now. However, there is a special procedure to move the soldering iron. This procedure is more unlikely to cause defects. It is soldering of an IC terminal. (DIP-IC,SOP-IC and QFP-IC) Look at the photograph in [Fig. 3-3]. Slide the soldering iron slowly. It is important to do it carefully. However, this work requires skills and cannot be recommended for a beginner. Do this work after acquiring the skills.

[Fig. 3-3] QFP-IC and its terminals

(3-4) Scattering of "solder balls" (dispersion)

[Fig. 3-4] is a state of the flux of the wire solder used in the soldering iron work. Certain solder has one core and the other has three cores. The flux disperses in the soldering iron work. However, the level of the dispersion decreases with wire solder having three cores compared with that having one core.

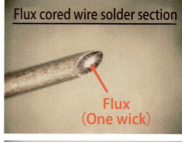
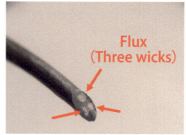

[Fig. 3-4] Wire solder section

3 Manual soldering: Various know-how

Certainly, flux disperses. How does solder become? Is there force to fly a solder ball of the metal at the temperature of the soldering iron? There is a doubt about **"dispersion of the solder ball"** though the flux disperses at the instant soldering iron temperature (about 350℃). After soldering iron work, the generation of a 'solder ball' is sometimes observed on the PWB. There is also an idea that the solder disperses with the flux. But, flux is not seen near the solder ball. The solder ball has been generated alone. Does solder really disperse? The test result is shown on page 45.

《 Solder dispersion test in soldering iron work. 》
Purpose: Confirm presence of dispersion of flux and solder during the work
What to prepare ⋯1. One PWB
     2. Two A3 white paper forms
     3. Soldering iron and flux cored wire solder
How to test ⋯Make a 10×10 mm hole in the center of an A3 form, and put it on the PWB.
  Test A ⋯Fix the soldering iron in the hole, and add wire soldering one after another.
  Test B ⋯Move the soldering iron up and down, and right and left, and add wire soldering in the similar manner. [Fig. 3-5]
After A and B are tested, check dispersion and dispersed items with a loupe or a magnifying glass.

[Fig. 3-5] Solder dispersion test

《 Solder dispersion test results 》
[Fig. 3-6] is a result of test A. Solder balls did not disperse though a lot of flux dispersed.
In test B, a soldering iron was moved. As a result, the dispersion of flux was confirmed as in test A. In addition, the dispersion of solder

45

To the world of defect-free soldering

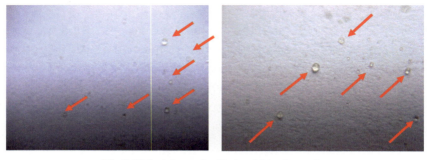

[Fig.3-6] Test A result, State of flux dispersion

balls was confirmed as shown in [Fig. 3-7], too. However, there was no flux near the dispersing solder balls. It is said that solder will disperse together with flux, but this seems to be wrong. The above-mentioned test results show that solder doesn't disperse as long as a soldering iron is not moved while flux disperses. Test B results show that chopsticks are put in and taken out of the glass containing water, dispersing drops of water. The melt solder is liquid like water. The drop of water doesn't disperse as long as chopsticks are not moved. Solder flies if a soldering iron is moved, and it doesn't fly if it is not moved. In short, it is up to a worker's skill level.

Solder doesn't disperse and the worker disperses it.

Therefore, it is possible to eliminate the problem by giving enough training and guidance to workers. The problematic work that disperses solder is that a worker moves a soldering iron more than necessary. **The role of a soldering iron is to raise the temperature of the metal to be soldered.** Apply the soldering iron to the metal and just wait until the temperature rises.

[Fig.3-7] Solder ball

3 Manual soldering: Various know-how

Relax and coffee time (4)

Flux dispersion and solder ball dispersion

Many tools to prevent flux of wire solder from dispersing are commercially available. The typical one is a "V ditch processing machine". This tool has both merits and demerits, is not easy to manage on the site, and is often left in the shelf of the factory though the idea of this tool is great. Flux adheres to teeth of the tool that is being used, making the V ditch gradually deeper, and the tool stops moving at the end. It is necessary to improve the tool although it is useful for preventing the target flux from dispersing. Moreover, it is also effective to use three-core flux as shown in the lower photograph for preventing the flux from dispersing. Various processing tools are commercially available to prevent this kind of flux dispersion. They are effective to a certain extent to prevent flux from dispersing, but not effective to prevent solder ball dispersion at all.

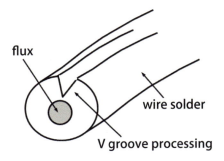

V-grooved wire solder

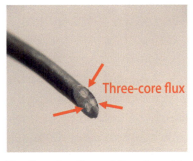

Flux three locations of the wire solder

(3-5) Method of repairing defective bridge between IC leads

A defect is as shown in [Fig. 3-8]. There are three methods of repairing this defective bridge as follows. However, the best repair method is No.3.

1. Repair method 1 is to forcibly wipe the bridge off with a soldering iron.
2. Repair method 2 is to apply the soldering iron to the bridge and to add the flux cored wire solder.
3. Repair method 3 is to apply the flux liquid to the brush, place the brush on the bridge part and apply the flux liquid. Then, apply there a soldering iron.

[Fig. 3-8] IC lead, bridging

(It is necessary to avoid repair method 1 or 2)

As for method 1, solder in the bridge part doesn't disappear only with one wipe. It is necessary to wipe the same part as two or three times off with the soldering iron. When the same part is wiped off many times, a pattern of the printed circuit board is peeled off. This repair method is against the "principle of soldering". The flux that remains in the bridge will disappear after gradually evaporating when the soldering iron is applied to the bridge many times. Then, the surface tension of the solder becomes larger and the bridge does not go away. This repair method is wrong.

Method 2, which is to add wire solder for repair, is seen in great many factories. In this case, only the flux is used, the melted solder will only be wasted by attaching to the tip of the soldering iron. With flux in the wire solder, the surface tension of solder is reduced. This method is possible with wire solder nearby and can be easily used by a repairer. However, one problem is why adding solder further to the bridge already having a large amount of solder. Solder is wasteed.

Method 3 is the best that solves the problem of method 2. And this is a repair method using an idea, "reducing the surface tension of

3 Manual soldering: Various know-how

the molten solder", which is the role of flux. The contents are shown in [Fig. 3-9].

It is necessary to always prepare flux liquid in a repair work site. The flux liquid is indispensable in the site like a soldering iron. Refer to explanations **i)** to **v)** in [Fig. 3-9].

《 **This is the method of repairing a bridge** 》

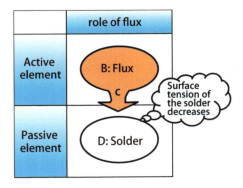

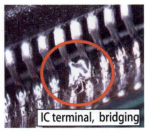

1. Apply flux. (Apply flux liquid to the bridge with a brush pen)
2. Apply a soldering iron to the bridge.

The defective bridge of **i)** is thought to be a bridge shape of **ii)**. ([Fig. 3-9]) The surface tension becomes smaller only by spreading the flux liquid in the bridge with the soldering iron, thus becoming **iii)**. In addition, if **ii)** and **iii)** are repeated, it becomes **iv)** and finally **v)**, and the repair is completed. This procedure is very simple only in accordance with the principle of a magic material, "The role of flux".

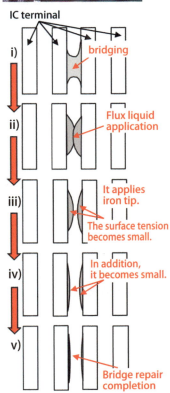

[Fig. 3-9] Method of repairing bridge where role of flux was considered.

To the world of defect-free soldering

And, this method can be used to repair all defective bridges including a bridge between IC terminals. Because wire solder is not additionally used, it is not wasted. However, use flux of the RMA type. The flux of the RMA type has comparatively small active force and does not require a printed circuit to be washed.

(3-6) Method of repairing soldering horn

[Fig. 3-10] is a sample of a defective soldering horn. A beginner puts the tip of the soldering iron to the corner and tries to remove the solder by force. In other words, it is to remove the soldering horn only with the soldering iron. It is not possible to remove it completely with this work. Applying the soldering iron only results in evaporating flux. Then, the surface tension grows and the soldering horn cannot be removed. (It is necessary to reduce the surface tension of the solder)

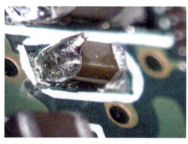

[Fig. 3-10] Solder protrusion icicle

It is against the 'principle of flux'. The correct repair is to spread flux liquid on the soldering horn with a brush pen and to apply the soldering iron there. Then, the horn completely disappears. This is an application of the role of the flux studied in the above-mentioned chapter to the site work. (Flux reduces surface tension) It is exactly a magic material.

(3-7) Replacing chip parts

There are three methods for this. The most recommendable is "Method of chip replacement···3". The reason is that it is unlikely to cause chip part damage or pattern pad peeling off as much as possible.

Method of chip replacement ··· 1

It is a method of detaching and installing a chip with one

3 Manual soldering: Various know-how

soldering iron. It is recommended to work with the knife-shaped iron tip as shown in [Fig. 3-11]. However, this work is possible only by an experienced expert. It takes time for a beginner to detach and install the chip only with a soldering iron. A defect, such as pattern peeling off or a chip crack, occurs as a result of struggling. This work cannot be recommended for a beginner.

[Fig. 3-11] The iron tip like the knife type

Method of chip replacement ··· 2

Remove a defective chip with two soldering irons as shown in [Fig. 3-12] and then solder quickly a new chip with two soldering irons. This work is possible by a person who is not necessarily expert but has some experience. However, on the repairing site at the factory, the site where two soldering irons are prepared for one repair person is few. It is recommendable to have two soldering irons for chip replacement. Moreover, if "magic material" flux is spread there before removing a chip, the work proceeds surprisingly quickly.

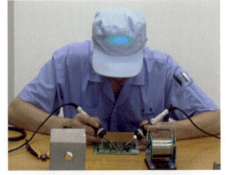

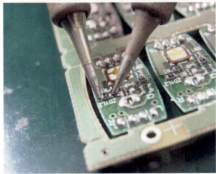

[Fig. 3-12] Work with two soldering irons

51

To the world of defect-free soldering

Method of chip replacement ... 3

The most recommendable is the tweezers type soldering iron as shown in [Fig. 3-13]. This soldering iron is designed for chip replacement and is very convenient. Some production sites have these soldering irons. However, they are not used while being left in the interiors of the shelves in the sites. The reason is that it is relatively easy to remove the chip but difficult to install it. Actually, the chip adheres to either iron tip and cannot be installed on the PWB well. This is thought to be the reason why it is rarely used. However, this soldering iron is very convenient and quality if this work is streamlined in various manners and the worker acquires the knack.

[Fig. 3-14] shows one devised example. Suppress chip parts with a thin bamboo skewer for installation so that they do not move. In addition, an experienced expert can install the chip skillfully even without the bamboo skewer. Soldering without a quality problem can be achieved if cream solder is spread on the pad when the chip is installed. This method...3 is very useful. It is recommended to acquire and use the skill knowhow as much as possible.

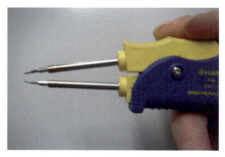

[Fig. 3-13] The soldering iron of tweezers type
Made from HAKKO CORPORATION

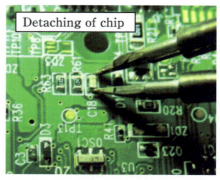

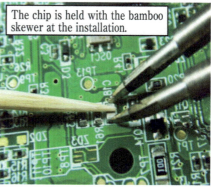

[Fig. 3-14] Example of devising work

3 Manual soldering: Various know-how

Relax and coffee time (5)

Casserole and soldering

"Casserole" and "soldering" seem to be very similar. As shown in the following figure and photograph, the ingredients of the casserole correspond to the metal to be soldered, and the gas stove corresponds to the soldering iron. There is no problem even if the foodstuff (for instance, leek and spinach, etc.) that can boil soon may be put in the pot later. However, the problem is the hard foodstuff (for instance, aroid) that doesn't cook easily. It will be pierced with chopsticks even if the potato is put in the pan from early time. Even if it cooks, it will be checked if it is ready to eat. Pay attention to the state of cooking of a hard foodstuff. Also pay attention to "temperature rise in a big metal" for soldering. Therefore, "hard foodstuff" is the same as "big metal". When a hard foodstuff becomes soft, it is time to enjoy a delicious pot dish. Avoid "cold solder".

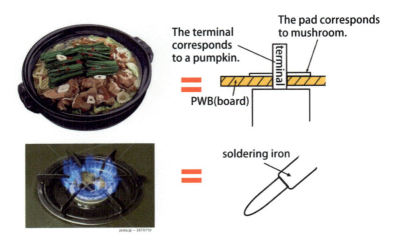

To the world of defect-free soldering

4   Measures against defective SMT reflow soldering

    The importance of the role of active element A heat (temperature) **a** in the soldering work has been explained up to now. A Heat (temperature) **a** similarly plays an important role to "**Soldering**" in auto soldering devices such as flow solder baths and reflow furnaces. Do not forget this A even with the automatic soldering system. Defects as shown in [Fig. 4-1] occur on the **SMT** (note) reflow soldering site every day. However, if the role of heat (temperature) **a** is understood and correct measures are taken against defects, it is possible to eliminate such defects.

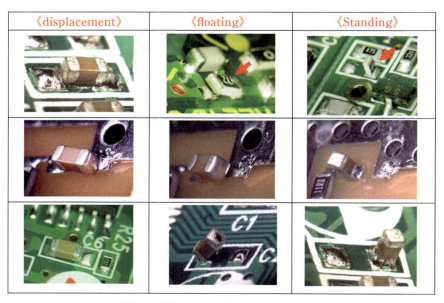

[Fig. 4-1] Examples of defective samples

    'Displacement', 'Floating', 'Standing (Manhattan)' have been put together. Because even if three defect names are different, their causes are the same. 'Displacement' if the influence of a defect is small. 'Floating' if the influence is bigger, and 'Standing (Manhattan)' if the influence is even more serious.

    [note] **SMT** stands for Surface Mounting Technology.

4 Measures against defective SMT reflow soldering

(4-1) Chip 'Displacement', 'Floating' and 'Standing'
(1) Cause of defects

Look at chips **A** and **B** ([Fig. 4-2]). After reflow, chip **B** was soldered without trouble. However, **A** has 'Chip floating'. While **B** was soldered in the reflow furnace without trouble, **A** has 'Chip floating'. The cause is "force" of surface tension of melted cream solder (left side of the chip). While **B** was soldered in the reflow furnace without trouble, **A** has 'Chip floating'. The distance

[Fig. 4-2] chip components good & defect

between chips **A** and **B** is 3mm or less. The reflow furnace seems to have become a magician. It is really mysterious. The cause of 'chip floating' is that the cream solder in both electrodes did not melt at the same time. Chip **B** melted at the same time. Therefore, chip **B** was soldered without trouble. Whenever the cream solder melts, the force called surface tension is generated. (Refer to [Fig. 4-3])

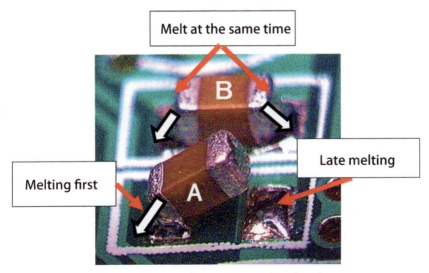

[Fig. 4-3] the surface tension (Pulling force)

To the world of defect-free soldering

In the case of B, force had been generated at the same time in both electrodes, so there was no floating defect as shown in [Fig. 4-3]. In the case of A, force was generated earlier than the right side because the left side melted earlier. The chip was floated. There was floating caused by surface tension (moment) on the left side. There was already no chip electrode on the pad even if the right side had melted afterwards. Why did the cream solder on the right of A melt later? There is a difference in size of the board pattern lands of both electrode parts even in an about 2mm-chip. The cream solder of both doesn't melt at the same time. The cream solder of the small land melts first, and chip standing is generated by surface tension generated at that time.

In a reflow furnace, there is inevitably a temperature difference when the temperature is inappropriately set. See [Fig. 4-4 and 5] of the board that has come out from the reflow furnace. There is a temperature difference even between approximately 1mm adjacent melt/unmelt pads under a certain condition. Melted lands (5 points) are small in [Fig. 4-4], and the only portion that did not melt is a big earth pattern land.

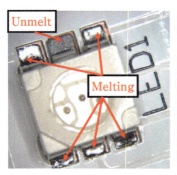

[Fig. 4-4] Melting, Unmelt

[Fig. 4-5] Melting, Unmelt

4 Measures against defective SMT reflow soldering

Why does a temperature difference (ΔT) occur?
1. Caused by the difference in size of the land (pad).
2. Caused by the temperature setting in the reflow furnace. (Inappropriate setting)
3. Caused by the difference in size of parts installed on the PWB.

With regard to 3, as parts on the board have already been decided, It is very difficult to improve them. Pay attention to 1 and 2. In the reflow furnace, the temperature of the small pattern (land) rises quickly than the larger land. Therefore, the cream solder on the small pattern pad will melt earlier. Moreover, each land of the board advances with a temperature difference to the preheating zone and the reflow zone when the temperature setting in the reflow furnace is not correct. And, there is a difference in soldering timing of each land even though the solder melts.

Therefore, the causes of chip 'displacement', 'floating', and 'standing' are as follows.

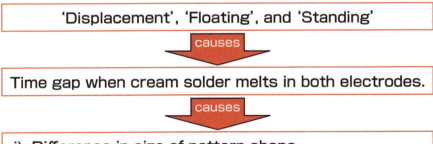

'Displacement', 'Floating', and 'Standing'
↓ causes
Time gap when cream solder melts in both electrodes.
↓ causes
i) Difference in size of pattern shape (Value of thermal capacity)
ii) Problem with temperature setting in reflow furnace (Heating is not uniform)

i)···The temperature of the small pattern rises quickly in the reflow furnace, and that of the big pattern rises more slowly. The measure is to eliminate the temperature difference.

To the world of defect-free soldering

ii)⋯There is a temperature fluctuate in each land on the board when the temperature setting is inappropriate in the reflow furnace. It is necessary to change the temperature setting in the reflow furnace to solve this problem. (The change in the temperature setting will be explained in more detail later) It is knowhow to keep the temperature rise of each land of the board the same. Moreover, it is knowhow to melt the cream solder at the same time. Explanations are given about measures i) and ii) in detail as follows.

(Note) It is also thought that the cause of chip 'displacement', 'floating', and 'standing' is "mount displacement" in the parts mounting process. However, there is a big doubt in this idea. Certainly, such a defect possibly occurs when the chip is greatly displaced from the pad portion in the mounting process. With regard to the printed circuit board that caused the "mount displacement" of square chips, the soldering state was investigated in the reflow furnace. The result is explained in the photographs in pages 77-79 onwards. It is soldered in a correct position without trouble with the self-alignment effect of the cream solder. **The self-alignment effect** is surface tension when the cream solder melts. Please refer.

(2) Countermeasures

Measures against these defects are to "**melt the cream solder of both electrodes of the chip at the same time**". There are two concrete measures. One is measures that can be taken in the technical department (pattern design department) and the other is those that can be taken in the production site as shown below.

1) Change a large pattern into smaller. (pattern design section)
2) Review the temperature profile in the reflow furnace and change to the "uniform heating setting" that melts big and small patterns at the same time. (production site)

4 Measures against defective SMT reflow soldering

Of course, it is impossible to make the large pattern (ex. earth pattern) smaller. However, it is possible to reduce thermal capacity only in the pad portion to be soldered among the large patterns. It is necessary to pay attention only to a necessary pad portion and make the shapes of the pads of both electrodes as same as possible. Then, the cream solder of both electrodes melts at the same time. **(Measure 1))**

Of course, the cream solder doesn't melt in the range of preheating below the melting point (217°C) (hereinafter, "preheating") in the reflow furnace. Therefore, both electrodes are heated in this preheating area as uniformly as possible. Afterwards, they advance to this heating area without temperature fluctuation and the cream solder melts at the same time. **(Measure 2))**

1) Make the large pattern smaller (Knowhow at the pattern design stage)

Chip parts shown in [Fig. 4-6] are soldered without trouble. However, if such a large pattern is left untouched, temperature difference is caused with both electrodes in chip parts, such a defect as 'displacement', 'floating', or 'Manhattan effect' of the chip will occur some time.

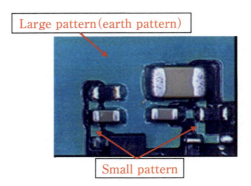

[Fig. 4-6] Example of big and small pattern

To the world of defect-free soldering

It is necessary to eliminate a size difference of the PWB pattern (pad). Change the big earth pattern side as is shown in [Fig. 4-7]. And, raise the temperature quickly in the reflow furnace. It takes longer to raise the temperature of pattern pad **X** in the figure compared with **Y**. Therefore, cream solder of **Y** melts earlier than that of **X**, and 'chip standing (Manhattan)' etc. occurs due to the surface tension of the cream solder of **Y**. **Z** becomes a small pattern (pad) when **X** is made like **Z**, the temperatures rise at the same time in **Z** and **Y**, and the cream solder melts at the same time and the surface tension works at the same time, not causing a defect. (Refer to [Fig. 4-7])

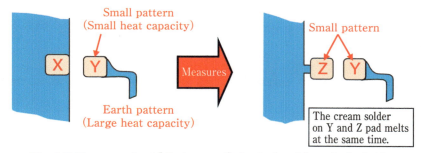

[Fig. 4-7] Measures about "displacement", floating" and "standing" of chips

《 State of soldering in measure patterns 》

The temperature rises in a small pattern quickly in the reflow furnace, and the cream solder on that melts immediately.

It is necessary to raise the temperature of the whole big earth pattern, but the temperature does not rise soon enough.

The temperature rises equally for the pattern lands of both electrodes of the square chip in the reflow furnace and the cream solder melts at the same time if the shape of the big

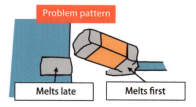

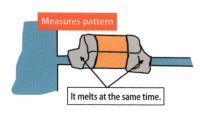

[Fig. 4-8] Measures pattern

4 Measures against defective SMT reflow soldering

earth pattern is changed as shown in [Fig. 4-7]. (Refer to [Fig. 4-8])

Compare a big continent of the United States and a small continent of the Florida peninsula. These two continents are the continuation of the land. In terms of the shape, the improvement pattern in the PWB looks like the Florida peninsula in the United States. In this book, such a pattern is called the "**Florida pattern**".

Concrete examples of the "Florida pattern"

Concrete examples of the 'Florida pattern' are shown as below. [Fig. 4-9] These concrete examples are decided and executed by the designer of the Engineering Department. Quality improves with fewer defects in the production site if the designer executes the improvement plan of the **Florida pattern**.

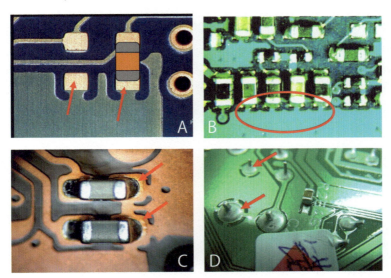

[Fig. 4-9] Example of Florida peninsula pattern

The Florida Peninsula land of **B** is a board for a cellular phone. The designer's intention is sufficiently reflected there. A high density board like the cellular phone is likely to follow the path of defect -> unrepairable -> discard, so the Florida Peninsula pattern is a really big improvement.

To the world of defect-free soldering

《 Florida peninsula and soldering improvement 》

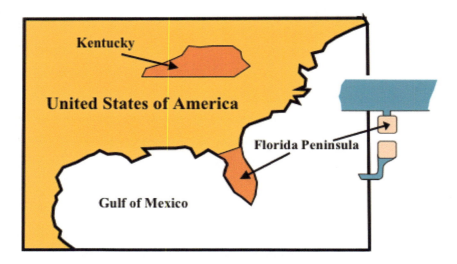

It takes time to raise the temperature inside the continent of the United States (eg Kentucky). The Florida peninsula is a land-based peninsula on the same continent, so temperatures are faster than Kentucky. You can raise it.

4 Measures against defective SMT reflow soldering

2) Review of temperature setting (Knowhow of the Manufacturing Division)

Next is measures to 2) review temperature setting. The composition of SMT (surface mounting) production site is basically as shown in [Fig. 4-10]. The board enters the cream solder printing process, and odd-shaped parts such as chip parts and ICs are mounted afterwards, followed by soldering in 3. Reflow process.

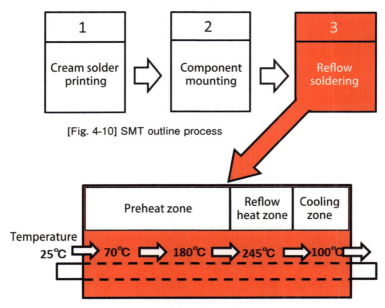

[Fig. 4-10] SMT outline process

[Fig. 4-11] Reflow furnace

The temperature setting in the reflow furnace is to raise the temperature little by little as shown in [Fig. 4-12] in the preheating area. However, a defect such as "Chip standing" occurs depending on how to raise the temperature. Therefore, it is necessary to do it while keeping **heating homogeneously**.

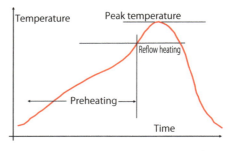

[Fig. 4-12] Example of temperature profile

63

To the world of defect-free soldering

Importance of "homogeneous heating"

There is inevitably a temperature difference (ΔT) on the board in the reflow furnace. Unfortunately, it is impossible to make this ΔT with 0. The temperature setting (**uniform heating**) in the furnace to be brought close to 0 becomes important. There are the following three causes of the temperature difference (ΔT).

1. In the reflow furnace, because the temperature rises in analog form, the temperature rises from the front to center and back of the board.
2. It is not easy to raise the temperature in the pad near the big odd-shaped parts (connector and IC, etc.).
3. A temperature difference is caused by the size of the copper foil pattern.

(A fan is installed in the furnace as a countermeasure so as to avoid a temperature difference as much as possible)

These are thought to be causes of ΔT. Measures (knowhow) to reduce a temperature difference become very important to avoid defective soldering. To melt the cream solder as simultaneously as possible even if there is a big or small pattern that is not the Florida Peninsula pattern as shown in [Fig. 4-13], it is necessary to set the temperature to heat uniformly at the stage of preheating. 'Chip standing' happened frequently because the temperature of the board shown in [Fig. 4-14] was not set to uniform heating at the stage of preheating.

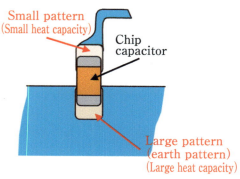

[Fig. 4-13] Example of large and small pattern

4 Measures against defective SMT reflow soldering

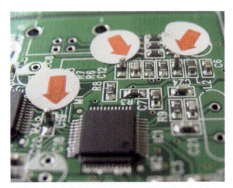

[Fig. 4-14] Chip capacitor, Frequent chip standing defect

The temperature setting in the reflow furnace was changed to uniform heating (measures against chip standing shown in [Fig. 4-14]). Then, the defects of "chip standing" decreased dramatically. (Refer to "Relax and coffee time (7)").

《 Method of setting temperature to "Uniform heating" 》

There are two types of temperature setting methods **A** and **B** in the temperature profile of the furnace as showing in [Fig. 4-15]. Which type has fewer defects?

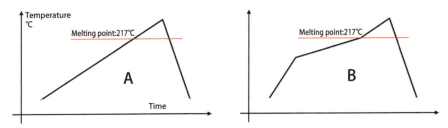

[Fig. 4-15] Temperature profile, Two types

As a conclusion, the profile of **B** has overwhelmingly fewer defects. The reason is that the "uniform heating" state is established in the preheating zone up to melting point (217℃) in **B** rather than **A**. The cream solder of both electrodes of the chip can be melted more simultaneously if "uniform heating" can be done. Measures against

65

To the world of defect-free soldering

'chip floating' and 'chip standing' are to use the temperature profile of **B**.

A ··· The temperature is gradually raised from normal temperature (25℃) and reaches the melting point. In this case, the cream solder of the small pattern melts first, causing a defect because it reaches the melting point with the state of temperature difference (ΔT) between the big and small patterns.

B ··· The temperature is higher than **A** in the preheating zone. The cream solder doesn't melt in the range below the melting point even if the preheating temperature is raised. The temperature is higher also in the big pattern than in **A**. The pattern is heated with the temperature difference (ΔT) being small, and solder of the big and small patterns melts more simultaneously. **(The temperature difference of both electrodes of the chip has become smaller)** (However, another defect occurs if the temperature of the preheating zone is raised too much. Be careful)

From page 67, the advantage of the temperature profile of **B** will be explained in detail. Moreover, explanations will be given on knowhow when a set value is changed from **A** to **B**.

66

4 Measures against defective SMT reflow soldering

Examples of setting profile temperature for "uniform heating"

The reflow furnace in zone 10 is composed of ten furnaces, and each has a heater that works as a heat source whose temperature can be set. One example of the temperature profile (for lead-free solder) in zone 10 is shown in [Fig. 4-16]. The temperature profile in **A** to raise the preheating temperature gradually and melt the cream solder at the melting point (217℃) looks ideal, but 'chip floating' and 'chip standing' happen frequently. Because the temperature distribution on the board doesn't become "uniform heating" in this temperature profile, and the board temperature advances up to the melting point with the temperature difference being kept.

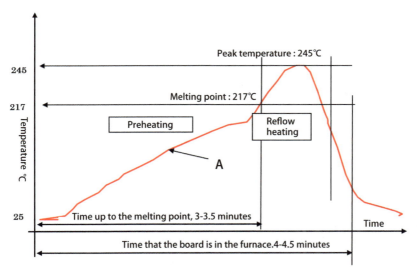

[Fig. 4-16] Temperature profile of A

67

To the world of defect-free soldering

Temperature setting of profile A

The temperature profile setting value of **A** is provisionally as shown in [Fig. 4-17]. The temperature is gradually raised, and this looks ideal but actually causes a defect.

(10zone)

zone No.	1	2	3	4	5	6	7	8	9	10
Temperature setting value	130℃	150	155	160	165	175	175	225	260	265

[Fig. 4-17] Temperature setting value of A

(Note) There are two heaters, heat sources for the furnace, in the upper and lower portions of respective zones, and it is usual to set the same temperature in both portions. A set value of the above-mentioned heater is displayed as one for both of the upper and lower portions.

Temperature setting to change A to B

Three profiles of **A**, **B**, and **C** are shown in [Fig. 4-18]. The defective soldering decreases dramatically if the temperature setting is changed to become a profile like **B** because **A** has the problem. Because the rise in temperature in **B** is thanks to more "uniform heating" in the area of the preheating compared with **A**. One doubt is raised here. There is an expectation that the defect will decrease further if the temperature is preset for the profile of **C**, which enhances the performance of **B** that is supposed to be good. However, although **C** is interesting in terms of an idea, it causes another defect and is not recommendable. The activation with the flux in the cream solder works earlier for **C,** and then phenomena of the strong oxidation of the base metal and solder in the furnace cause a defect..

4 Measures against defective SMT reflow soldering

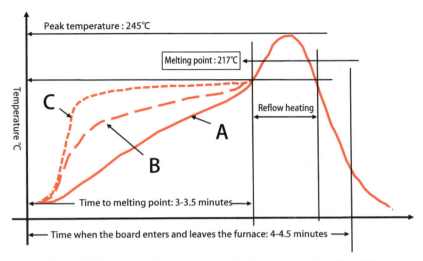

[Fig. 4-18] Example of temperature profile (Improvement profile of B)

For instance, set the temperature profile value of **A** as shown in [Fig. 4-19] now. Change the temperature setting as shown in red to heat **A** more uniformly. Then, it becomes the profile of **B**.

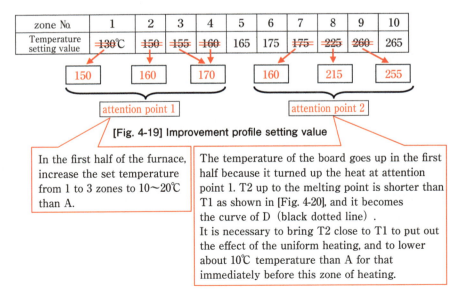

[Fig. 4-19] Improvement profile setting value

In the first half of the furnace, increase the set temperature from 1 to 3 zones to 10~20℃ than A.

The temperature of the board goes up in the first half because it turned up the heat at attention point 1. T2 up to the melting point is shorter than T1 as shown in [Fig. 4-20], and it becomes the curve of D (black dotted line).
It is necessary to bring T2 close to T1 to put out the effect of the uniform heating, and to lower about 10℃ temperature than A for that immediately before this zone of heating.

To the world of defect-free soldering

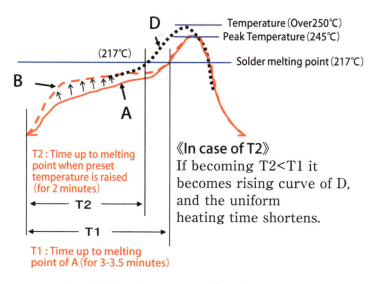

[Fig. 4-20] Knowhow to make B's profile

Countermeasures against chip 'displacement', 'floating', and 'standing' are:
1. Florida Peninsula pattern considered in the pattern design stage.
2. Temperature setting to heat uniformly by the Production Department (Review of temperature profile).

These two different countermeasures have the same intention.

---

1. Measures against the 'Florida Peninsula pattern'
2. have the same intention as those against the 'temperature profile review'.

   The measures are to melt the cream solder at the same time in both electrodes of the chip parts.

4 Measures against defective SMT reflow soldering

Moreover, the defective soldering is corrected by setting the temperature of **B** for uniform heating as shown in the right side of [Fig. 4-21]. It is strongly recommended to review the temperature profile.

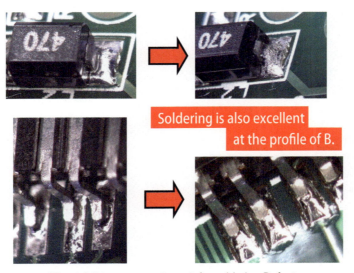

[Fig. 4-21] Improvement result for soldering Defects

To the world of defect-free soldering

Improvement examples by setting temperature change

It has been explained that there are fewer defects in the temperature setting of **B** than that of **A** in the temperature profile. Here is a concrete example.

Story in a certain factory ......

It was the factory of SMT where high density boards were produced. One day, the leader on the site was very worried and shouted, "There are a lot of 'chip standing' cases and the line is totally disrupted! A defect rate exceeded 50%! Help!" The defect rate data had been taken in this site every hour. The defect occurrence rates were as shown in the right graph. Surprisingly, the defect rate reached 65% at 12 o'clock! As a result of many investigations on site, the pattern was not the 'Florida Peninsula'. The profile of the reflow furnace showed the temperature setting of **A** after all. When it was changed to **B** and data was taken after 15 o'clock, the defect rate fell down to 0.7%.

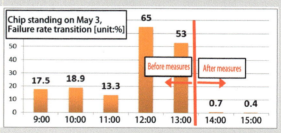

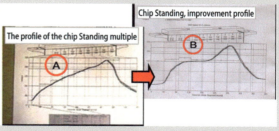

It was obvious that the temperature setting of **B** was correct although it was still not perfect.

72

4 Measures against defective SMT reflow soldering

(Note 1) Is the temperature profile correctly measured?

The temperature profile becomes a monitor of the board temperature and the soldering temperature under production, and an important document to check the "temperature at that time" when a defect occurs. Despite being an important document, a measurement sample (board) is sometimes defective, resulting in an inaccurate temperature survey in the furnace. Pay attention to the following two major problems in the production site.

(**Problem 1**) ⋯ How to make a thermocouple

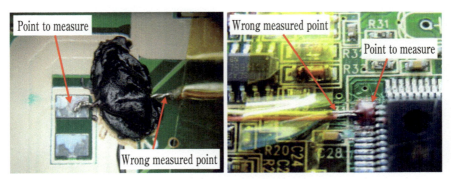

[Fig. 4-22] Where is the measurement point?

[Fig. 4-22] is a bad example showing that the part to be measured and the part actually measured are different. Taking a temperature profile with such a sample results in an irrelevant temperature profile with inaccurate measurement temperature and/or the measurement of the atmosphere of the furnace. The thermocouple measures the temperature at the first intersection of the two main lines (bare conductors). Therefore, if the intersection is made as shown in [Fig. 4-23], it is necessary to separate the other bare conductors. (Bad example [Fig. 4-24])

To the world of defect-free soldering

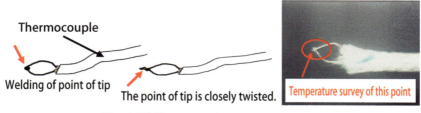

[Fig. 4-23] Thermocouple, processing method

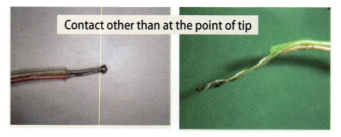

[Fig. 4-24] Thermocouple sample with problem

(**Problem 2**) ⋯ How to fix a thermocouple ([Fig. 4-25])

When fixing the tip of a thermocouple (intersection) to a target pad, use an adhesive and a heatproof tape for the method. If the amount of the adhesive is excessive and/or too many tapes are piled up, heat conduction becomes insufficient and the measured temperature becomes inaccurate.

[Fig. 4-25] There are a lot of adhesives and tapes and the problem.

4 Measures against defective SMT reflow soldering

[Fig. 4-26] shows a comparison with the profile between a defective sample and a correct sample. When a large amount of adhesive is used, the curve of the peak is round (R is large). It is sharpened in a correct sample (R is small), which is obvious at the sight of the profile.

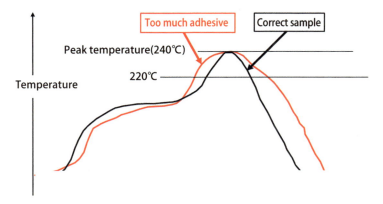

[Fig. 4-26] Comparison of peak tip shapes

(Note 2) What happens if the chip is displaced from the mount?

The defective soldering (chip displacement, etc.) as shown in [Fig. 4-27] sometimes occurs in the production site. Usually, the cause is a mounter machine, and the countermeasure becomes a "replacement of the nozzle tips of the mounter". Such cause and measure are always identified every day. However, even if the chip has shifted before the reflow, wrong measures are taken and lead to a repeated defect if it is judged that the mounter is the cause. After the chips before soldering are moved by intention, how become after the reflow? Look at the following test photograph.

(However, the reflow temperature profile was done in the correct temperature setting as shown in **B** on page 69. The defect as shown in [Fig. 4-27] possibly occurs in the temperature setting of **A**.) About the all test photographs on pages 77-79, the left side shows those whose parts were displaced by intention and the right side shows

To the world of defect-free soldering

those soldered after passing the reflow furnace. It is a result of an "effect of self-alignment" by the surface tension of the cream solder itself.

[Fig. 4-27] chip displacement

4   Measures against defective SMT reflow soldering

Sidewise displacement test of 1608 types (chip resistors)

《 Move it by intention 》   《 State of soldering after reflow 》

To the world of defect-free soldering

Diagonal shift test of 1608 types (chip resistors)

《 Move it by intention 》　　　　《 State of soldering after reflow 》

4 Measures against defective SMT reflow soldering

Sidewise displacement test of 1608 types (chip capacitors)

《 Move it by intention 》　　《 State of soldering after reflow 》

To the world of defect-free soldering

It is found from the above-mentioned test results that, if solder in both electrodes are melted at the same time with a correct temperature profile, soldering is carried out without trouble. They are soldered without trouble with the "effect of self-alignment".

(4-2) 'Solder ball' defect

A 'solder ball' generated in a high density board of SMT is a defect to cope with. There is no guarantee that this solder ball does not move in the customer of the post process or the market. If the solder ball falls between IC terminals on the board, it causes a short circuit, and the product itself becomes a big problem in the market. Therefore, this 'solder ball' has to be totally eliminated on a high density board. Before taking countermeasures, divide this defective name into the following two for better understanding. Because the causes of the defects are different though they come from the same **"solder ball"**.

The two defective names are ⋯

(1) 'Side ball', generated on the side (just beside) of the chip. Refer to [Fig. 4-28]. The cause is one.
(2) 'Solder ball', generated in various places on the board. There are many causes.

Photographs of respective defects are shown below.

[Fig. 4-28] Side ball

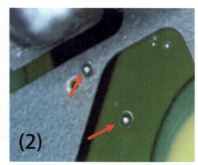
[Fig. 4-29] Solder ball

4　Measures against defective SMT reflow soldering

(1) 'Side ball' cause and measures

A 'side ball' is generated on the side (just beside) of the chip as shown in [Fig. 4-28]. The cream solder disperses by no means in the reflow furnace. (However, flux disperses) Next, the causes of the generation are explained. The cream solder is printed on the board (PWB), and chip parts are mounted on the cream solder and then enter the reflow furnace. There are the preheating process and the heating process in the reflow furnace, and the cream solder on the board causes a sagging symptom (preliminary heating sagging) in the preheating process as shown in [Fig. 4-31] in the reflow furnace. This preliminary heating sagging becomes the 'side ball' by the following 1 and 2.

1. The melting cream solder causes preheating sagging and, in addition, the cream solder extends on the board by the capillary action between the board surface and the chip surface.
2. The spread cream solder is cut with the corner chip of 90°.

With regard to 1 ⋯⋯

The symptom of cream solder sagging is more or less caused in the preheating zone in the reflow furnace. In three photographs in the following [Fig. 4-30], the center shows the state of preheating sagging, the left is when the cream solder is printed, and the right is when the solder is melted with this heating at temperature higher than the melting point, comes out of the furnace and hardens.

(The three photographs are of the same board, and a dotted line circle has the same size. The center is bigger than the circle. There is also a state of bridge.)

[Fig. 4-31] shows how much sagging occurs in the cream solder under the square chip. The cream solder that melted and widely spread as sagging in the reflow furnace was not able to gather on the pad. As a result, the defect of a "side ball" occurred on the side of the chip as shown in the photograph.

To the world of defect-free soldering

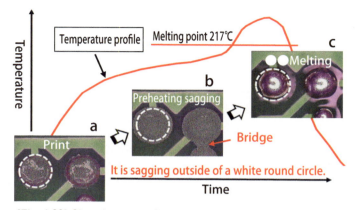

[Fig. 4-30] Changing state of cream solder along temperature profile

[Fig. 4-31] Sagging state and side ball

With regard to 2 ······

Here is an unusual example of the board (Refer to [Fig. 4-32]). It is a hybrid board of a square chip and a round chip.

Look at the two kinds of chip parts. The side ball is generated in the square chip (red arrow) and not in the round chip. The side ball is also seen in mini-diode (rectangular). However, the side ball is not seen

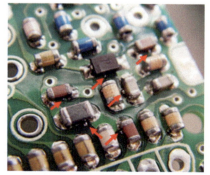

[Fig. 4-32] The mixed-loading board of a square chip and a round chip.

4 Measures against defective SMT reflow soldering

on the board although a lot of round chip (Melf-type chip) parts are installed there. These phenomena are very interesting.

Difference between square and round

About the square chip and the round chip, the cream solder similarly cause the sagging symptom in both lands in the preheating zone. Why is the side ball generated only in the square chip? It is due to the shapes of the square and the round. The corner of the square chip is 90°. In contrast, the round chip is circle. (Refer to [Fig. 4-33])

The cream solder that causes preheat sagging starts gathering on the pad when the temperature becomes more than the melting point (217℃) of heating. However, as part of the sagged solder is cut by the corner of the square chip, the rest cannot gather in a land and is left at the side of the chip. This is the side ball defect. However, the section of the round chip is circular, there is no obstacle to it. Then, all the sagged cream solders gathers on the pad without cutting, and the side ball is not generated.

The above is an explanation of **2**.

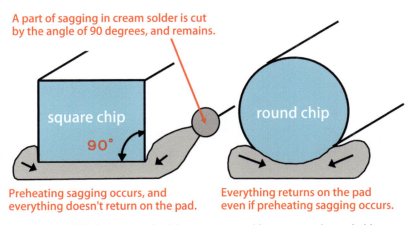

[Fig. 4-33] Sagging and solder movement with square and round chip

To the world of defect-free soldering

Measures against the side ball

If the real cause of the side ball is found, it is not difficult to take countermeasures. The following two points can be implemented. Even if only one of the two points is executed, the effect is obvious.

(1) Improvement and change of metal mask

The hole shape of the metal mask can be changed as improvement in the production site. (However, the hole area of the metal mask is the same as that of the conventional one, and the amount of solder doesn't change) An example of an improved mask hole is shown as follows. If shape **2** in [Fig. 4-34] is changed to **3** or **4**, the side ball can be improved.

These improvements are to prevent the cream solder from spreading to the side and/or below the chip components. (Additional)These improvements are really big for a defective side ball. (Additional)(I changed the hole shape of the metal mask in the past, so that a defect of a side ball was totally eliminated in the main board of a notebook computer)

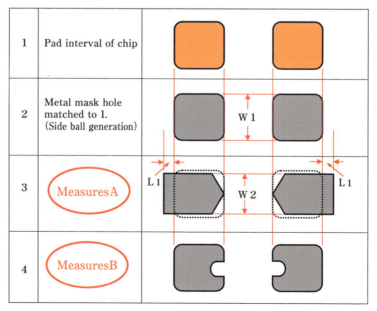

[Fig. 4-34] Metal mask hole for side ball measures.

4 Measures against defective SMT reflow soldering

Measures A and B shown in [Fig. 4-34]

1 shown in [Fig. 4-34] shows a pattern land shape of the printed circuit board. The hole shape of the metal mask is usually adjusted to that of the pad of **2**. However, when preheating sagging becomes serious, a 'side ball' is generated. As a countermeasure, make it to the home base shape of baseball to keep the side ball from sagging on the side of the square chip as shown in measure A of **3**. (However, the hole area has to be the same as that of **2**) Measure A: W1-W2=0.2 mm Both sides have been narrowed respectively by 0.1mm. Widely expand them even if being narrowed. (L1=0.2 mm) Solder gathers on the pad after melting without problem. The hole shape of measure B of **4** is often seen in the production site. This hole shape is unlikely to cause capillary action and holds preheating sagging in check. **3** is recommendable rather than **4**.

Recently, there is also cream solder that hardly causes preheating sagging thanks to the maker's efforts. Therefore, 'side balls' dramatically decrease with such cream solder that seems to be very effective if used. The difference between the one used on site and the conventional one is surprising. Therefore, 'side balls' dramatically decrease with such cream solder that seems to be very effective if used. However, it is not actually easy to change cream solder in a site using SMT, although it would be better if the cream solder could be changed easily. There are problems with quality, cost and customer approval, so it is not possible to change it soon. One realistic idea will be to rely on the improvement (this change in metal mask hole) that can be made in the SMT site until the solder can be changed.

To the world of defect-free soldering

Relax and coffee time (8) **Cream solder and soft ice cream**

As a child, anyone who has bought a soft cream in the park at the summer heat season will be in everyone. The soft serve ice cream is beautiful in shape for a while. However, the temperature goes up and soft serve ice cream starts sagging under the scorching sun. Haven't you been troubled with melted soft serve ice cream dropping on your hand? The cream solder in the reflow furnace also cannot endure and causes sagging like soft serve ice cream in the environment where the temperature has risen in the preheating zone. (Preheating sagging)

Sagging of soft ice cream

4 Measures against defective SMT reflow soldering

Relax and coffee time (9)

Relax and coffee time about low temperature solder (9)

The cream solder is classified to three kinds by melt temperature into "high temperature solder", "usual solder frequently used", and "low temperature solder". In "low temperature solder", Bi (bismuth) is contained and causes serious preheating sagging. This cream solder causes various defects in a high density board and is improper. Pay moderate attention to this when using it. Refer to the sample of the preheating sagging situation of the low temperature solder (containing Bi). (Photograph below)

The cream solder gathers in the center when the heating process starts, although there is serious sagging. However, a very thin bridge (micro-bridge) is possibly generated, which is difficult to visually check, when the temperature setting in the reflow furnace is inappropriate. Pay attention to temperature setting.

State change of cream solder containing Bi

Cream solder application

Preheating sagging

Soldering state after main heating

87

To the world of defect-free soldering

(2) 'Solder ball' cause and measures

A 'solder ball' has been generated randomly on the board. The defect shown in [Fig. 4-29] is supposedly caused not by splash of the cream solder in the reflow furnace but by another reason. It seems that power that can splash or fly cream solder is not generated anywhere in the environment where temperature rises gradually from normal temperature. There are a lot of causes of 'solder balls' as shown below, and those measures are also different respectively. Without executing them at the same time, it is impossible to eliminate the 'solder balls'.

《 Causes of solder balls 》
1. Print misalignment of cream solder
2. Defective automatic cleaning of printer
3. Applying a tape on the back of metal mask
   (Tentative measures)
4. Insufficient board washing after misprinting in the cream solder
5. A worker's error (adjustment of a mounted part) immediately before the reflow furnace
6. Preheating sagging of cream solder

Respective measures
1. Print misalignment of cream solder

The print accuracy cannot be secured when the cream solder is printed, especially in a high density board such as QFP-IC, causing misalignment. In such a case, serious preheating sagging occurs in the preheating process of the reflow furnace, and a part of the cream solder that cannot gather on the pad in this heating process remains on the board as a ball. The countermeasure is to improve print accuracy. When beginning the production, the person in charge on site needs to visually check print accuracy of the board thoroughly. Moreover, another countermeasure is to change to cream solder that is unlikely to cause preheating sagging. Recently, boards have been of higher density. It is important to adopt the cream solder that is unlikely

4   Measures against defective SMT reflow soldering

to cause preheating sagging as a measure against a "solder ball" in such a situation.

2. Defective automatic cleaning of printer

Cream solder adheres to the other side of the metal mask hole when continuous printing is carried out several times. If this state is left, the amount of solder decided by contact printing increases, thereby becoming gap printing. After the reflow, there are such defects as a 'solder ball' and a 'bridge'. Due to these problems, an automatic cleaning device is usually installed in the printer now. The cream solder that adheres to the other side of the metal mask is automatically wiped off when printing is carried out several times. However, although there is no problem with the purpose of this device, there is no guarantee that it can remove the solder that adheres by 100% as it is automatic. The removed solder sometimes adheres to the other places of the mask when solder on the other side is wiped off. Minute cream solder that adheres to the other side of the metal mask is adhered on the board that is placed next. The board to which a small amount of cream solder adheres advances to the mounting process and the reflow process, and a 'solder ball' is generated on the board when it comes out from the reflow furnace. The above-mentioned explanations are given in A to D of [Fig. 4-35]. Measures against such a 'solder ball' are to do manual cleaning (an operator wipes off with a cloth that contains alcohol) after automatic cleaning, thereby completely removing the cream solder that remains on the other side.

To the world of defect-free soldering

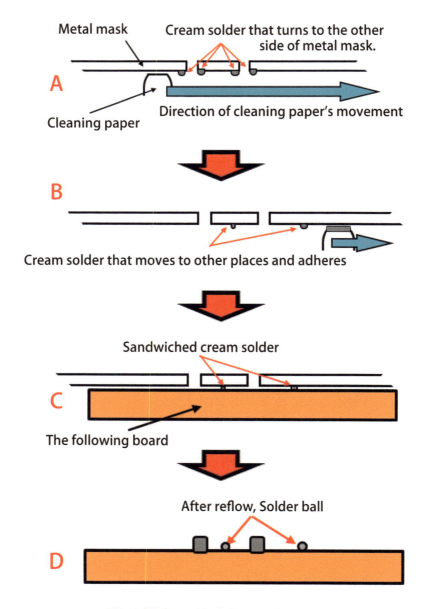

[Fig. 4-35] One solder ball generation example

4　Measures against defective SMT reflow soldering

3. Applying a tape on the back of metal mask

There is an unexpected blind spot in automatic cleaning of 2. A further blind spot is to paste a tape to the other side of the metal mask. To increase the solder quantity when only an 'insufficient amount of solder' was generated, there is a measure that can be immediately and provisionally implemented by a technical member and the person in charge on the SMT production site. The measure is to apply a tape to the other side of the metal mask. This is just a provisional measure. For instance, if a **0.02**mm tape is applied to the 0.1mm-thick the metal mask, the amount of solder theoretically increases by 20%. This can be satisfactorily sufficient as a measure against an 'insufficient amount of solder'. And, as this tape application is immediately executable in the production site, all technical members handling SMT have this expertise. However, there is a blind spot also in this tape pasting work. Naturally, if the tape is applied, there is a difference of **0.02**mm on the other side of the metal mask. If the other side of the metal mask is cleaned automatically with a difference, cream solder that is wiped off is caught to the difference and only a smaller amount of solder remains. (Refer. to [Fig. 4-36]) Afterwards, a 'solder ball' is generated as explained in 2. The only countermeasure against a 'solder ball' that is generated as such is to stop applying a tape on the other side of the metal mask. The following three measures are recommended to increase the solder quantity while stopping applying the tape.

i) Replace the polyurethane rubber squeegee with the metal squeegee.
ii) Review printing pressure.
iii) Review the metal mask hole size (Enlarge a little).

[Fig. 4-36] Adhesion near the step of the tape, cream solder

To the world of defect-free soldering

4. Insufficient board washing after misprinting in the cream solder

Look at the board shown in [Fig. 4-37] on the right. This looks like a new board to be input to the production process.

However, this board had been once printed, resulting in a printing error. After the error, all old cream solder on the board were washed down in alcohol. But, a small amount of cream solder is found to remain

[Fig. 4-37] Print mistake

with a microscope. ([Fig. 4-38]) A printing error doesn't actually occur frequently. The printing error has been sometimes caused by a setting error of a new printer after a replacement and/or a little displacement of the board. Because the error rarely occurs, its handling is in fact not emphasized in a lot of production sites.

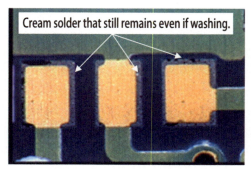

[Fig. 4-38] It remains still.

When a board as shown in [Fig. 4-38] is input to an SMT line again, a 'solder ball' is generated after the reflow. The measure is to wash off the majority of cream solder manually during alcohol washing. Then, there is no problem if it is further washed off with a supersonic wave washing machine.

5. A worker's error (adjustment of a mounted part) immediately before the reflow furnace

In some SMT production sites, an operator (as well as an inspector) is seen between the parts-mounting process and the reflow furnace. The operator is adjusting the displacement of a mounted part with a thin bamboo skewer. While adjusting, he/she is causing another defect. However, at this time, he/she is causing even another defect.

4 Measures against defective SMT reflow soldering

If a mounted part is displaced, it is only necessary to adjust the setting of the mounter machine. It is better for the operator to avoid moving parts for adjustment. Because an operator's pressure is not constant though the pressure of the mounter machine (Z axis) is constant. The cream solder on the pad is mashed up and spread when pushed strongly. This causes the generation of a 'solder ball'. Recent mounters have dramatically improved accuracy so that it is unnecessary to adjust them manually. However, there is displacement caused by abrasion of a nozzle that sucks parts. Displaced boards continue if the nozzle is not immediately replaced at that time. As a result, the operator takes action. Moreover, the many recent parts become complex and miniaturized in shape. These are really difficult for the production site to handle. The manufactures are requested to improve them as early as possible. Look at the 'solder ball' shown in [Fig. 4-39]. It is the ball that was generated due to poor handling of an operator.

The operator mistakenly touched the cream solder on the board while working and put it on another part. Especially, a 'solder ball' in the vicinity of the edge of the board (vicinity of the ear of the board) is caused by an error of the operator, as he/she holds the edge of the board by the hand for working.

[Fig. 4-39] The worker's mistake, Solder ball

6. Preheating sagging of cream solder

A solder ball is generated when the symptom of "1. Print misalignment of cream solder" is serious.

Therefore, when beginning production, it is important to visually confirm print accuracy of cream solder. Never forget that this operation is the basic in the production site.

To the world of defect-free soldering

(4-3) 'Less solder' cause and measures

[Fig. 4-40] shows defective 'less solder', which means a small amount of cream solder on the pad of QFP-IC. It was because a foreign matter or hard cream solder adhered to the relevant hole of the metal mask. An appropriate quantity of cream solder was not printed. Of course, if the foreign matter etc. is not removed, less solder continuously occurs. (Refer to [Fig. 4-41])

In a high density board, even an extremely small amount of foreign matter causes less solder like this. The Production Division has to thoroughly manage the printer. For instance, the cover of the printer has to be closed during production. ([Fig. 4-44] shows problematic operations)

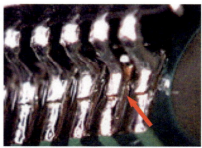
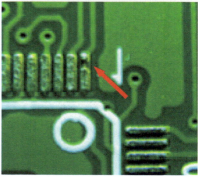

[Fig. 4-40] 'Less solder'

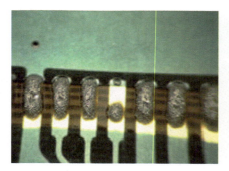

[Fig. 4-41] 'Less solder' at the same place is continuous if the foreign body of the metal mask is not removed.

Cause of 'less solder'
Foreign matter adheres to metal mask
Types of foreign matters
(1) Adhesion of old cream solder
(2) Garbage adhering to board
(3) Small dust adhesion in SMT production site

The biggest number of 'less solder' cases is seen in an IC (especially QFP-IC) terminal pad as shown in [Fig. 4-40]. Naturally, the metal mask hole becomes smaller in accordance with a narrow pitch of 0.3mm. In the current state of the smaller metal mask hole, it is better to avoid deciding that the cause of 'less solder' is cream solder. There is an anxiety that the cream solder does not easily come off the hole of the metal mask in the current state that the hole becomes small. 'Less solder' relapses later if a cause of the defect is cream solder. Particles of the cream solder are globular. All the particles are globular now while some of them were not globular in the past. The particles are globular and have a diameter of about 20μm. It is 0.02 in terms of mm.

The particles having a diameter of 0.02mm go through the metal mask hole without trouble. However, there is 'less solder' in the first two pieces at the beginning of the production. This is because flux of the cream solder (solvent etc.) doesn't fit the surface and section of the mask enough. An appropriate quantity of cream solder is printed from the third piece. [Fig. 4-42] shows a cream solder particle.

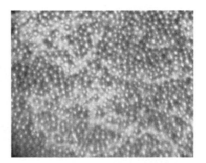

[Fig. 4-42] Cream solder particle.

To the world of defect-free soldering

(1) Old cream solder adhesion and the countermeasures.

After completion of production, a used metal mask has to be immediately washed to remove the adhering cream solder. Washing includes automatic washing and manual washing. After the automatic washing, it is necessary to check fouling visually or with a loupe. Cream solder particles possibly adhere to the section if the metal mask is stored without being able to be washed. Look at [Fig. 4-43].

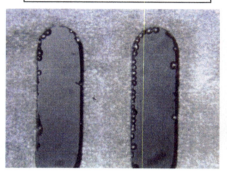
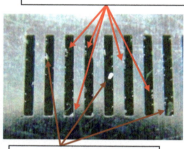

[Fig. 4-43] After washing the metal mask Situation of IC terminal hole

An inspection after the metal mask is washed is the duty of the person in charge of the site. It is also more effective to establish check items to prevent 'less solder' from occurring and to patrol in the process periodically.

(2) Measures against foreign matter that adheres to board

This foreign matter is brought in by the board maker. The board maker cuts the board and makes a hole on the board with a drill. Of course, these procedures are implemented with automatic equipment, and powder of the board adheres to the board at this time. If the board to which powder of the board adhered is input to the production process of SMT, the powder causes less solder on the metal mask. The board maker is requested to take the countermeasures, but this

4 Measures against defective SMT reflow soldering

is not enough. It is necessary to remove the powder on site before inputting the board to a printer (Blow off the powder by air or remove it with an adhesive tape).

(3) Small dust adhesion in SMT production site and the countermeasures

The SMT production site is maintained in an orderly manner seemingly without any dust or foreign matter. However, when it is observed carefully, there are small foreign substances, etc. In such an environment, small foreign matters and dust adhere to the mask when the cover of the printer is opened as shown in [Fig. 4-44]. During production, make sure to close the cover of the printer.

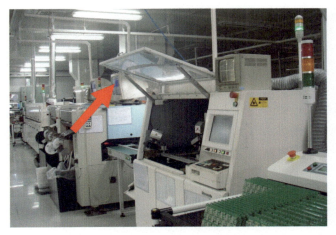

[Fig. 4-44] Productuon site with the cover of the printer opened.

To the world of defect-free soldering

(4-4) 'Bridge' cause and measures

There are mainly three causes of a 'bridge' defect that occurs in the SMT reflow process.

(1) When the cream solder was printed, the amount of solder was too much.
(2) The amount of solder was too much due to deformation of a metal mask.
(3) As the temperature setting in the reflow furnace was inappropriate, cream solder concentrated on one point after melting.

(1) When the cream solder was printed, the amount of solder was too much.

It is usual to use contact print in the current high density board so as not to set the gap between a metal mask and a board. [Fig. 4-45] shows that the terminal IC pad is appropriately printed. However, cream solder comes to the other side of the mask little by little as explained in "Automatic cleaning device" on page 88 when this contact print continues. If the solder that adheres to the back side of the metal mask is not wiped

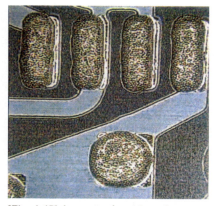

[Fig. 4-45] Amount of solder of correct contacting print

off and printing continues, it is gap print rather than contact print. As a result, the board moves to the next process in the state of an excessive amount of solder.

Refer to [Fig. 4-46], [Fig. 4-47] and [Fig. 4-48]. And, it results in printing with the excessive amount of solder as shown in [Fig. 4-48]. This excessive amount of solder coming out from the reflow furnace is a 'bridge' defect. The measures against this are to completely wipe off

4 Measures against defective SMT reflow soldering

the cream solder on the back side of the mask and return it into the state of contact print.

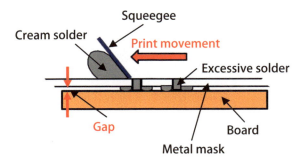

[Fig. 4-46] Figure of gap print

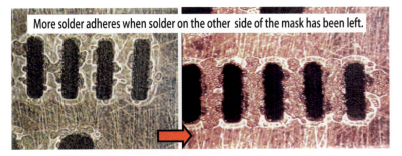

[Fig. 4-47] It adheres to the other side of the metal mask.

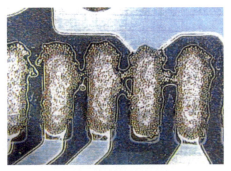

[Fig. 4-48] State of bridge already to print excessive solder

99

To the world of defect-free soldering

(2) The amount of solder was too much due to deformation of a metal mask.

It is necessary to handle the metal mask with care. However, unfortunately, metal masks are not handled or stored carefully in some production sites. In particular, in a site where high density boards are produced, once a mask is deformed even a little due to improper handling, it is impossible to return to the original shape, possibly causing a fatal defect. One example is a bridge of QFP-IC on a high density board.

Moreover, there was a 'bridge' defect in a certain IC terminal. The defect rate was about 60%. The cause was found to be a metal mask deformation as a result of an investigation. A recent metal mask has board thickness of 0.1-0.15mm as shown in A of [Fig. 4-49]. Although contact print was carried out as shown in B, the deformed metal mask had gap print in a certain portion as shown in C. And, 'bridge' continues to be seen in the same part. It is necessary to produce new metal masks as a countermeasure, or the defect continues to occur.

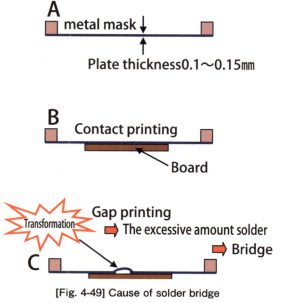

[Fig. 4-49] Cause of solder bridge

4 Measures against defective SMT reflow soldering

(3) As the temperature setting in the reflow furnace was inappropriate, cream solder concentrated on one point after heating and melting.

This kind of 'bridge' defect sometimes occurs in a narrow pitch IC, especially in QFP-IC. As a principle, melting solder moves to the place of higher temperature. Therefore, cream solder that caused preheating sagging tends to move to a pad of higher temperature. It is important to eliminate a temperature gap on a board and heat it uniformly as a countermeasure against a bridge. There is a 'bridge' on a pad to which a large amount of melted cream solder moves.

Therefore, the countermeasures are to review the temperature setting in the reflow furnace and to subject the board to uniform heating. The methods have already been explained after page 64. Refer to 《 Method of setting temperature to "Uniform heating" 》. It is shown again in [Fig. 4-50] for reference. The measure is to change the temperature profile setting of **A** to that of **B**.

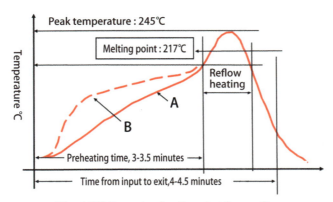

[Fig. 4-50] Example of uniform heating profile

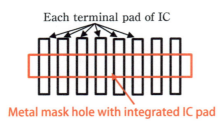

101

To the world of defect-free soldering

5   Measures against defects of 'Flow (DIP) soldering'

The flow solder bath was used to solder "lead insertion parts" as shown in [Fig. 5-1] at first. Then, PWB boards have been of higher density and further miniaturized. The boards were advanced to two-side from one- side. Insertion parts have been mounted on the upper surface and surface-mounted components have been installed on the lower surface for flow soldering. The surface-mounted components on the lower surface are temporarily fixed with an adhesive so as not to fall.

The outline framework of the flow solder bath is as shown in [Fig. 5-2]. The board is put on a conveyer of a soldering furnace and enters therein, and next, is applied flux (liquid) at under side of the board. Then, the board goes through the preheating (about 100℃) process in the heater. Then, it enters melting flow solder bath (250 - 260℃) and is soldered.

[Fig. 5-1] Lead inserting parts and PWB

There are two baths. The first half is a jet bath (the first bath) that forcibly applies a large amount solder of melted to the board, and the latter half is a stationary bath (the second bath) to finish up soldering. The board soldered in the solder bath is cooled afterwards and discharged from the flow bath.

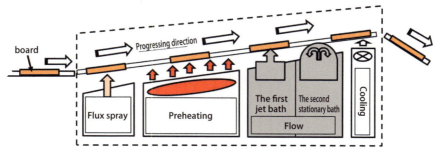

[Fig. 5-2] The outline composition of the flow soldering equipment

5 Measures against defects of 'Flow (DIP) soldering'

The flow-soldered boards have been of higher density and miniaturized, and use of SMD parts has caused even more defects on the production site. Defects such as 'bridge' and 'no solder (Solder has not adhered)' have occurred frequently on the production site. Causes of these defects and the knowhow of how to cope with them will be explained.

(5-1) 'Bridge' cause and measures
《 Examples of bridge defects 》

Most of defects caused by flow soldering are overwhelmingly 'bridges'. Narrow pitch parts (for example, IC and connector) have a pitch interval of the terminal of 2.5 to 3.0mm. The distance between such terminal pattern pads becomes 1mm or less, always causing concern about a bridge. Moreover, the amount of solder in the solder bath, a large amount of molten solder, is equal to infinity compared with a pattern pad with a small board. As a large amount of molten solder is applied to the board, a bridge is likely to occur as a matter of course.

[Fig. 5-3] the bridge of IC terminal   [Fig. 5-4] the bridge of SOP-IC

[Fig. 5-3] shows a "bridge" defect of insertion parts IC (DIP-IC of 2.5mm pitch), [Fig. 5-4] shows a "bridge" of surface-mount IC (SOP-IC), and [Fig. 5-5] shows a "bridge" of a connector terminal. These defects have respective causes, and it is easy to take measures to prevent the bridges from occurring as long as their true causes are found.

103

To the world of defect-free soldering

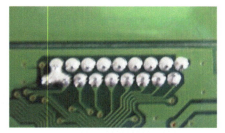

[Fig. 5-5] the bridge of the connector terminal.

《 Measures against 'bridge' 1 》

······**Knowhow of flow direction of the board at flow soldering.**

There is an important commitment (knowhow) to flow-solder narrow pitch parts (for example, IC or connector). The knowhow is to decide **the direction of the board** when it is passed through the solder bath. The number of occurrences of bridges was tested in a different direction. There are different boards **A** and **B**, in which boards equipped with narrow-pitch IC and connector are arranged as shown in [Fig. 5-6]. Which parts arrangement **A** or **B** has fewer 'bridges' against the flow direction? Test results are shown as follows. (No matter how many times this test is repeated, the result is almost the same.)

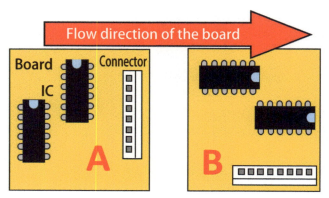

[Fig. 5-6] Which parts arrangement A or B does 'Bridge' decrease?

5 Measures against defects of 'Flow (DIP) soldering'

Test results of A and B (20 each)

	Number of bridge failure occurrences
A	38 points / 20 units
B	9 points / 20 units

DIP-IC (16PIN-IC, 2.5mm interval)

The number of bridge defects dramatically decreases when the board is flowed to the direction as shown in B. Far more 'bridges' occur in arrangement **A** than **B**. Therefore, it is necessary to arrange parts as shown in B at the stage of the pattern design, and the person in charge on the production site has to pay attention to the "direction of the board". **What the pattern designer of the Engineering Department needs to take care of first is to arrange narrow-pitch parts, such as IC and connectors, in a constant direction. Then, it is necessary to screen-print the arrow in the constant direction on the ear on the edge of the board. (Refer to** [Fig. 5-7]**)** If this knowhow is not kept, many 'bridge' defects occur, followed by an increasing number of corrections in the post-process, and the production site is likely to be confused.

And, there are considerably fewer "bridges" when the board is flowed to the direction of **B** than that of **A**. However, the number of bridges doesn't become 0. A "bridge" inevitably occurs at the end of the terminal as shown in [Fig. 5-4] and [Fig. 5-5]. The cause and measures of this 'bridge' will be explained as follows.

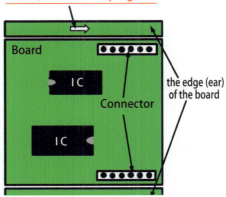

[Fig. 5-7] The arrow of flow direction

105

To the world of defect-free soldering

《 Analyses of cause of 'bridge' occurrence 》

Why does such a 'bridge' occur between the ends of terminals in the moving direction? The cause is analyzed as follows. Indeed, it is mysterious. Look at [Fig. 5-8]. The first and second IC terminals are just coming out from the molten solder bath. Naturally, 1 and 2 possibly become bridges due to narrow pitches.

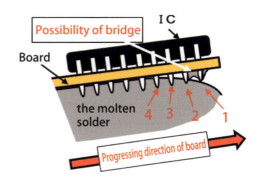

[Fig. 5-8] Cause of generation and the analysis about 'Bridge'

However, the board advances from the left to the right. The molten solder between 1 and 2 in the drawing immediately moves to 3 due to the terminal of 3. A bridge did not occur between 1 and 2 because the molten solder moved to 3. Likewise, the molten solder of 2 and 3 moved to 4, and a bridge did not occur in 2 or 3. The solder in which a bridge possibly occurs thus moves to the next terminal, and consequentially, a bridge does not occur. And, the molten solder comes to the last terminal. The molten solder in the last terminals causes a bridge as it cannot move anywhere. (Refer to [Fig. 5-5] and [Fig. 5-9]) The measure against this is to make space available for molten solder as solder cannot move and has remained in the last terminal. The samples of the measures are shown in [Fig. 5-10] and [Fig. 5-11]. These effective measures can be taken only at the stage of the design of the board pattern. The number of defects decreases sharply with these measures on the production site, leading

[Fig. 5-9] why? Last terminal bridge

5 Measures against defects of 'Flow (DIP) soldering'

to no confusion. At the tests of the above-mentioned A and B, 38 defects happened in A due to no space for the molten solder to move in all of the terminals.

《 Measure against 'bridge' 2 》
······Knowhow at pattern design stage

At the tests, a 'bridge' was not completely eliminated even if the parts arrangement of B. 9 bridge defects that occurred during the test are seen between the last pads. Therefore, these countermeasures are to make space available for redundant molten solder. A concrete example is shown in [Fig. 5-10]. It is very difficult to execute these countermeasures only on the production site. All boards that have come out from the flow soldering equipment are inspected on the production site. If there are defective goods, they are corrected and repaired with soldering iron. However, if these measures are executed at the pattern design stage, the number of defects becomes zero and the production site is not confused.

[Fig. 5-10] Bridge measures, Example of pattern (Solder puddle pad at the rear part)

Therefore, if one more pad had been installed in an IC in [Fig. 5-9] as shown in [Fig. 5-11], a 'bridge' would not have occurred. Other examples of the pad (red arrow) as countermeasures are shown below.

It is very effective to add the pad as shown in [Fig. 5-11] as countermeasure knowhow of the pattern design section.

To the world of defect-free soldering

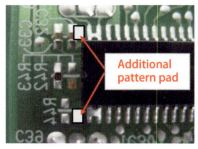
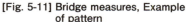
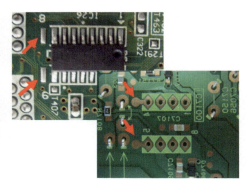

[Fig. 5-11] Bridge measures, Example of pattern

《 Other measures against bridge 》
······Knowhow at further pattern design stage

    Notebook computers, cellular phones and digital cameras, etc. have been further miniaturized. Therefore, printed circuit boards have been of higher density, and a pitch between pads has been narrower. Therefore, many 'bridges' have occurred on the production site. There are "bridges" even between in-between pads. (Refer to [Fig. 5-12]). There is a method of expanding space as much as possible between pattern pads of the terminal as measures against "bridges" Of course, it is impossible to expand the pitch interval itself, but it is possible to expand the pad between adjacent terminals in appearance. A narrow pattern with the resist pattern as shown in [Fig. 5-13] becomes a measure against the bridge. The board design section has to execute this measure. Strength is demanded at the terminal soldering part of the connector. Look at [Fig. 5-13]. There is possibly a concern about due to a little narrower horizontal solder filet, but by taking a lot of areas of the pad in the vertical direction of the amount and making it oval, the concern is thought to be eliminated.

5 Measures against defects of 'Flow (DIP) soldering'

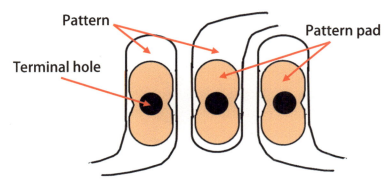

[Fig. 5-13] Bridge measures, Example of pattern

[Fig. 5-14] shows two examples as samples of measures against a bridge with the resist pattern.

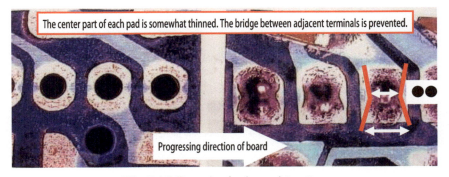

[Fig. 5-14] Example of using resist pattern

The center of each pad is thinned a little to prevent a bridge between adjacent terminals. It is impossible to execute such measures against a bridge on the production site. These measures have to be implemented at the pattern design stage. If the above-mentioned two countermeasures are executed at the pattern design stage, quality on the production site becomes steady for sure.

109

To the world of defect-free soldering

(5-2) Cause of 'No solder' and measures
《Examples of "No solder" defects》

The defect of 'No solder' occurs on the terminal pad of the transistor as shown in [Fig. 5-15] on the flow soldering site. As a large amount of molten solder is applied to the pad in the solder bath, it seems unthinkable that such a defect as "no solder" occurs. However, such a defect actually occurs every day.

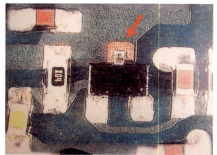
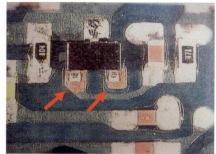

[Fig. 5-15] Defect example not adhere solder

Explanations will be given on why this defect of "no solder" occurs and measures against it.

《 Cause of 'No solder' and measures 》

Originally, mini molding transistor (minimo thereafter) parts are surface mount devices (SMD parts) and are used for SMT reflow soldering very frequently. However, due to densification and miniaturization, such minimo parts and chip parts, etc. have been fixed with an adhesive and flow-soldered. In such a situation, the cause of defects becomes obvious by looking at restrictions caused when the minimo parts-installed boards pass through the solder bath. It takes about 4-5 seconds for the board to pass over the solder bath shown

5 Measures against defects of 'Flow (DIP) soldering'

in [Fig. 5-16]. It will be necessary to solder in the short time without trouble. However, flux gas and air, etc. gather in the space (red arrow in [Fig. 5-17]) between the board side and the molding resin side of the minimo transistor (TR) in the molten solder bath. And the flux gas and air obstruct soldering in the space. As a result, 'No solder' occurs in 4-5 seconds. Of course, most transistor terminals are correctly soldered in the primary solder wave bath. However, when remaining gas is not totally discharged, this 'No solder' takes place. The chip electrode itself is a metal in the right square chip shown in [Fig. 5-16]. As molten solder is quickly applied to the pad, it is not affected by gas and such a defect as mentioned above doesn't occur.

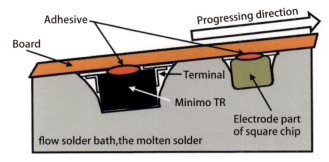

[Fig. 5-16] Parts in the solder bath

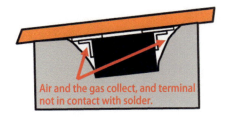

[Fig. 5-17] Minimo TR, air, and gas

To the world of defect-free soldering

《 Measures against 'No solder' 》
······Knowhow at pattern design stage

As the cause of the above-mentioned trouble was found, the following measures are to be taken.

1. Measure to remove gas and air
2. Measure to induce melted solder to the pad and to solder forcibly in a short time even when the gas and air cannot be removed sufficiently.

Measure of 1······Open a hole to drain gas of the board. (Measure at pattern design stage)
Refer to [Fig. 5-18].
Hole diameter: 0.8-1.0mm φ

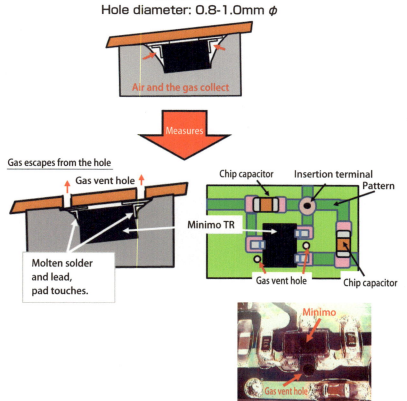

[Fig. 5-18] Example of measures for gas vent hole.

5 Measures against defects of 'Flow (DIP) soldering'

Measure of 2······Build a solder road (road of solder).(Measure at pattern design stage)

It is necessary to finish flow soldering in about 4-5 seconds successfully. Molten solder easily adheres to chips and the terminals of parts as a matter of course when the board flows to the soldering bath. A "road of solder" is built as shown in [Fig. 5-19] to guide the solder from the adhering part quickly, in a short time. (Measures at pattern design stage)

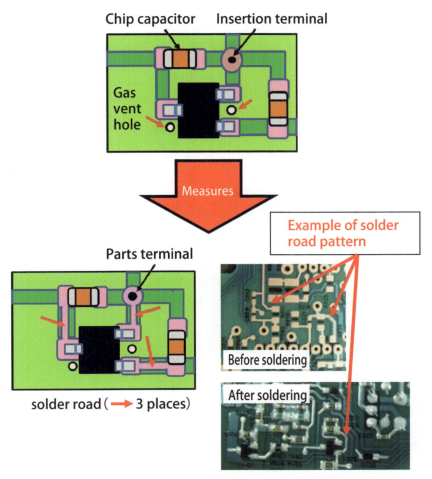

[Fig. 5-19] Example of measures for solder road

To the world of defect-free soldering

If the above-mentioned **two measures (1 and 2)** are taken, the 'no solder' defect is totally eliminated. The improvement example is shown in the following [Fig. 5-20] for reference.

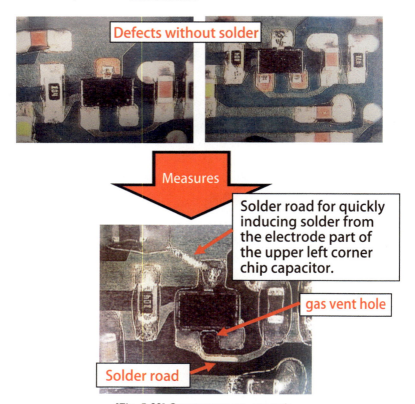

[Fig. 5-20] Countermeasure example

'Measures against soldering defects'

It is the responsibility of the people in the production site to take the preventive measures if a defect occurs. (On-site engineers and persons in charge of the Production Department and the Quality Management Division) In many cases, measures against these defects and the preventive measures are discussed and decided in a meeting room, etc. In such a case, trying to draw a conclusion hastily without investigating defective goods or grasping the situation on the

5 Measures against defects of 'Flow (DIP) soldering'

production site often results in irrelevant measures. It is surprising that a lot of conclusions made at meetings on quality are irrelevant as causes of defects. For example, they are "Parts are bad", "The board is bad" or "Solder is bad". In many cases, the causes are attributed to something else without investigating defective goods carefully. Appropriate measures cannot be taken if a true cause of the defect is not found, and the production site is confused. As a result, a defect is repeated. Most causes of defects exist on the production site. Our seniors said, "It is important to go to the site first of all when a defect occurs and look at a defective article carefully." This word is a royal road. Reality cannot be seen only at the discussions on quality in a meeting room. Giving as many as 4 or 5 causes of a defect at a meeting on quality means that a defective article is not investigated thoroughly. For example, many chip parts were suddenly damaged at a certain factory. The conclusion at the meeting on quality at that time was "Parts are bad". The person in charge requested the parts manufacturer to take countermeasures, but the defect occurred later again. As a result of a thorough investigation on the defective articles, the cause was found to be the manufacturing worker's handling of the board. As a result of changing the handling by the worker, the defect was totally eliminated. True causes of defective goods are found with a thorough investigation. It is recommendable to analyze a cause is analyzed with the above-mentioned stance.

Important points for finding a cause of a defect were mentioned in the first part of this book.

《 We understand it not because things are there but because we look at it closely. 》

Analyze a cause of a defect in such a manner.

To the world of defect-free soldering

6 | Soldering mystery collections (Who is the criminal?)

Examples of measures against defects on site

This chapter is a concrete reference material used for investigating causes of defects. A true cause can't be found easily. Time and tenacity are necessary to reach it at last. The cause lurks in an unexpected place. It is easy once the true cause is found. It is like a mystery until it is reached. It looks like the action of a detective who solves a difficult case. There are a lot of mysterious true stories. Very recommendable.

....

Clients troubled with a defect whose cause is uncertain visit me every day.

....

"Oh, I was surprised!"
A request for an investigation of a defect came also today.
Read through and use them on the production site.

---

1. Lines A and B: Why are there many customer complaints only about boards produced in Line A?
2. Suddenly, a 'chip crack' occurs.
3. Mystery of 'Print misalignment' on a flexible board.
4. Does the characteristic of solder change suddenly!?
5. Mystery of terminal IC "No solder adheres".
6. Mystery of production plant late at night.
7. Suddenly, 'board swelling' happens.
8. The display of the timer disappeared suddenly three years after buying!?
9. Suddenly, a pattern is lost.
10. Defective goods are full of treasures! (Important know-how hidden there)
11. Automatic driving, is it okay?

116

6 Soldering mystery collections (Who is the criminal?)

(Episode 1) Lines A and B: Why are there many customer complaints only about boards produced in Line A?

"The cause cannot be found easily. Not at all. Please help me!"

"What's wrong with you, panicked so much?"

"The same PWBs (printed wiring boards) are produced in lines A and B, but there are somehow a lot of customer complaints only about boards produced in Line A. I really have no idea."

A man of strong physique coming to consult now: He is an on-site plant manager at an SMT (surface mounting board) soldering production factory. He has a splendid beard suited for the physique. Speaking while patting the beard is his habit, but he seems to be so troubled as to forget about the boast.

I heard the story in detail.

The same PWBs are produced in lines A and B.

All parts and facilities used (printer, mounter, and reflow furnace) are the same, and the workers and the manager hold also the posts of both lines A and B in the same manner.

However, there are a lot of customer complaints about boards produced in only line A. There are few complaints about boards from line B.

The production manager had been perplexed with the problem, saying, "I can't understand at all why this phenomenon occurs".

The above is the plant manager's story.

This is certainly strange. However, there must be the truth on the production site.

"Plant manager! Let's go to the production site together!"

The quality manager, the plant manager, the Production Department director, and the persons in charge of the production, etc. who were attending a quality meeting, went to lines A and B on site.

Everyone checked the parts and equipment used in lines A and B. All of them were exactly the same.

It was impossible to find any difference between lines A and B

117

To the world of defect-free soldering

A line : Setting of temperature of reflow furnace

Zone No..	1	2	3	4	5	6	7	8	9	10
Temperature setting value	130℃	150	170	200	180	180	185	220	260	260

B line : Setting of temperature of reflow furnace

Zone No.	1	2	3	4	5	6	7	8	9	10
Temperature setting value	130℃	150	170	170	180	180	185	220	260	260

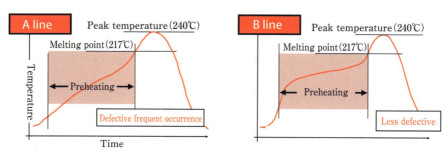

[Fig. 6-1] Temperature setting value of A and B line, and profile comparison of both.

even after an hour had passed, and everyone was at a loss.

After about two hours had passed, when a certain engineer compared the temperature profile material of two reflow furnaces(A and B), he found one mysterious phenomenon.

It was quite strange!

"What's the matter?"

The problem was found in the reflow furnace!

"Plant manager, there is a problem in the reflow furnace".

"The actual temperature curve of the profile in line A has been low although its temperature setting in Zone 4 is higher than that in line B".

"Don't you think this is against the principle?"

"Look at [Fig. 6-1]. The setting temperature in line A is 200℃, 30℃ higher than that in line B, 170℃, in terms of the setting of the fourth furnace in the preheating zone. However, the temperature of line A is lower than line B in the profile curve of the preheating zone".

6　Soldering mystery collections (Who is the criminal?)

"Don't you think this is against the principle?"

"Oh! It is certainly strange".

"The temperature is lower as a result, although the preset temperature is high".

"Let's check the inside of the reflow furnace in line A".

"Hey! Everyone! Take off the cover of the furnace!"

[Fig. 6-2] Heater line, disconnection !

The cover of the reflow furnace of line A was removed by all workers.

"Oh, what a surprise! The third heater line is disconnected!" ([Fig. 6-2])

"Oh! The third heater line has been disconnected!" (Refer to [Fig. 6-2])

"Really! Completely disconnected!"

"The temperature in the preheating zone did not rise as the third heater was disconnected. The temperature of the fourth heater was set abnormally high (200°C) for correction".

Heating was not uniform in each land on the board as the correction was made to heat rapidly within the preheating range. As a result, the many defects occurred in line A.

Everyone on the site understood the importance of uniformly heating in the reflow furnace.

"Thank you! The cause has been found".

"Repair the disconnected heater line to set the temperature correctly".

A month later

"Since then, we have repaired the disconnected line A and adjusted its temperature setting to the same as that in line B. Then,

To the world of defect-free soldering

the quality improved dramatically, and there is no complaint from customers". "Thank you very much".

"There are many defects of the temperature profile when heating is not uniform as seen in line A. It was good the problem was solved".

《 Episode 1: The End 》

> 《 Quality improvement person's comment from Episode 1 》
> Ladies and gentlemen, beware that temperature cannot be seen directly. When such a defect occurs, it is recommendable to doubt whether the temperature profile is set to uniform heating.

(Episode 2) Suddenly, a 'chip crack' occurs.

The door was intensely knocked.

One day, a man jumped into my room. The man with brown hair in white working clothes seemed to be in his twenties.

"I'm really upset!!".

"Oh, just calm down. What happened to you?"

He became calm little by little and began to talk about the story of his work.

"Suddenly, chip parts were broken. The damage also occurred on a board of a different production model".

"Chip parts are not broken every day but occasionally".

[Fig. 6-3] chip crack

"The damaged parts are chip resistors or chip capacitors, etc."

"I am troubled because my superior said that the parts were

6   Soldering mystery collections (Who is the criminal?)

damaged due to my poor board handling".

The damage of the chips isn't eliminated no matter how carefully we work. (Refer to [Fig. 6-3])

Apparently, he seems to get angry at his superior judging that the cause of the defect lies in his handling of a board.

"I understood your feelings well. First of all, let's survey the production site".

I went to the site with him, and he started working as usual.

I checked all the processes on the SMT production site, and there was no special problem. Also, there was nothing wrong with the site workers' board handling.

The cause was not able to be found on the day.

I went home on the day and I decided to investigate the cause on another day.

Moreover, when I came home with him on that day, he told me, "A 'chip crack' does not occur every day".

Results of the investigation on the day

1. Chip damage is not limited to certain parts. It is seen in chip capacitors and/or the chip resistors.

   (Because damage occurs in parts from different makers, they are not responsible for it)

2. There is no special problem with the workers' handling of boards.

3. A 'chip crack' does not happen every day but once or twice a week.

The defects were really mysterious, and it was impossible to grasp the cause at all.

I went to the site several days later again, and restarted an investigation.

To the world of defect-free soldering

After all, there was no problem in his handling of the boards. He treated them more politely than other workers.

"Isn't there any problem with the equipment?"

The equipment on the site started to be reviewed.

[Fig. 6-4] board rack

One board rack was in sight about an hour after the review had been started. ([Fig. 6-4])

"Please show all the board racks here!"

There is a loader where the board rack is put as shown in [Fig. 6-6] in the first process on the SMT production site.

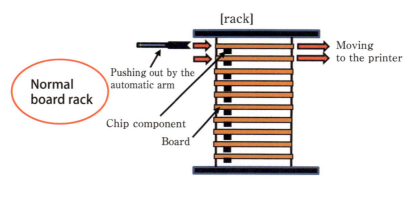

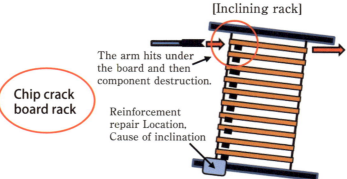

[Fig. 6-5] Inclining rack

122

6 Soldering mystery collections (Who is the criminal?)

There is the loader in front of the printer as shown in the figure.

First comes the loader in the process, the rack is put on the loader, and boards are pushed out by an automatic arm to the printer one by one.

Boards that have come out from the reflow furnace are collected in the rack again automatically at the end of the process.

The unloader is where the rack is put. ([Fig.6-6])

On the site, there was one rack with a problem!

"Look! It is a little inclined!"

There was one rack that had been reinforced and repaired.

The reinforcement method was that "packing tape" was wrapped around the place where the screw disappeared. ([Fig. 6-5])

"Look at this! This is it. It is a little inclined".

I assumed the movement of a board rack and an automatic arm with the loader and the unloader and forecast the cause of a "chip crack".

"Oh, really? Let's check it."

The plant manager put the rack in question on the loader and moved the arm slowly and manually.

"Oh! The arm doesn't hit the board but strikes the chip parts under the board!"

Moreover, a lot of splinters of the chip parts had been scattered on the floor of the loader.

Thus, the cause of chip crack completely found.

Refer to [Fig. 6-5] for detailed explanations of the cause.

The plant manager consented, and the site leader was satisfied with the above.

It was found that the engineer on the site had wrapped the rack in question with a tape for reinforcing the place where the fixation screw disappeared.

The rack was inclined a little by wrapping the tape. (Refer to [Fig. 6-5])

The crack of chip parts was caused by using the rack without noticing its inclination.

To the world of defect-free soldering

The plant manager said "I am sorry for doubting you".
This repaired rack was prohibited from using and was removed from the production site.

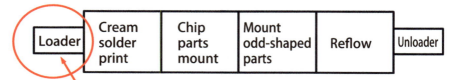

A Large number of broken pieces of chip parts on the floor under the loader

[Fig. 6-6] SMT process

At a later date
"I'd like to express my gratitude really. Thank you. I have been cleared of suspicion thanks to you, and there has not been any chip crack ever since".

"It was really good that the cause (true culprit) was found. The plant manager has understood that you are indeed serious with work".

《 Episode 2: The End 》

《 Quality improvement person's comment from Episode 2 》
Doesn't such a trouble occur at the equipment and tools around you? It is very important to check whether the machine works normally as the motion of the machine is decided by man.

6 Soldering mystery collections (Who is the criminal?)

(Episode 3) Mystery of 'Print misalignment' on flexible board

"The cause of the complaint of the other day was investigated, and my countermeasures were approved. As a result, I was appointed to the manager at the new factory".

The man who is proudly speaking while patting the beard is the plant manager who once requested me to investigate the cause of a trouble.

"It is wonderful that you are appointed to the manager at the new factory. Congratulations!"

"Ah. Thank you. But I'm not feeling very well".

"How come? What's the matter?"

"No, now, there isn't a big problem, but there is a concern".

We began to produce a lot of flexible board products at our factory now. It is the first experience for us.

I have long handled the conventional "copper -covered multi-layered boards" so far.

It is the first time for me to solder and commercialize flexible boards although I have known it since before.

In digital cameras and cellular phones, there are a lot of flexible boards. Recently, more and more people use them, and the production of high density boards has increase.

They are mainly produced in the new factory, aren't they? It's wonderful!

"That's for sure, but it's shameful for me as the plant manager to answer, 'The cause is not found' when there were some troubles like before. Therefore, I want to learn defects whose causes are difficult to find from you beforehand".

"I see. Knowhow different from the current one is certainly needed on the SMT production site of the flexible board. How about in this case?"

It is about a defect that occurred during cream solder print of mini connector 20PIN (0.4mm pitch) as shown in [Fig. 6-7]. The symptom is misalignment printed between pads, and its level is 0.2mm. Because

To the world of defect-free soldering

this actual thing becomes a defective bridge in the post-processing, it can't be sent to the next process. The print misalignment mysteriously occurs only in the part of this connector.

"Why does it happen?"

[Fig. 6-7] Print displacement

"Plant Manager, the story still continues".

Even more mysteriously, there are two kinds of misalignments on the right side A and the left side B for the pad. (Refer to [Fig. 6-8]) Moreover, the print misalignment doesn't occur in the pads of the chip capacitor, chip resistor, Mini TR, or DI on the same flexible board with the misalignment in the connector.

Why does the misalignment occur only in the connector despite the same flexible board used?

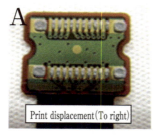 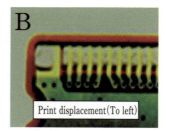

[Fig. 6-8] Displacement symptom

"It has become even more mysterious".

"Yes, workers on the site were also bewildered at that time. Then, I paid attention to two points".

"What are they?"

They are the following two points.

1. Why is the relation between the direction of the movement of the squeegee and the direction of the print displacement opposite?

2. Why does only the connector have a misalignment while there

6 Soldering mystery collections (Who is the criminal?)

is no misalignment of cream solder in chip parts, although both are on the same flexible board?

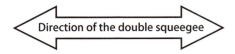

About 1

The squeegee of an automatic printer moves from the right to the left when printing, and then from the left to the right. (Double squeegee method)

The squeegee moved from the right to the left in A shown in [Fig. 6-9] (misaligned to the right side against the pad). And, it moved from the left to the right in B (misalignment on the left side).

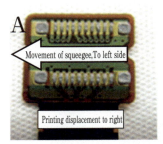
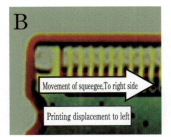

[Fig. 6-9] Cause investigation of displacement

"Normally, it is strange that the displacement occurs in the opposite direction to the movement of the squeegee".

See [Fig. 6-11] and [Fig. 6-12] for detailed illustration.

Pay attention to the **"reinforcing board"** shown in [Fig. 6-10] in the above-mentioned 1 and 2.

The reinforcing board on the other side of connector (t = 0.2mm)

[Fig. 6-10] the reinforcing board

127

To the world of defect-free soldering

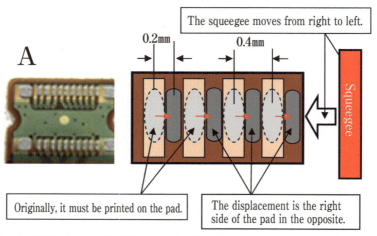

[Fig. 6-11] Displacement of the opposite direction against movement of squeegee (Shift on the right side)

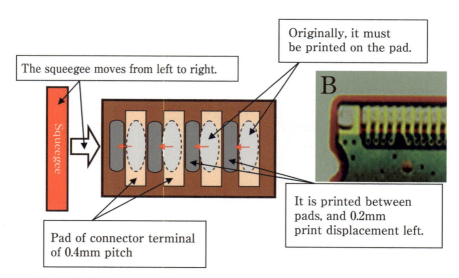

[Fig. 6-12] Displacement of the opposite direction against movement of squeegee (Shift on the left side)

6 Soldering mystery collections (Who is the criminal?)

It is necessary to explain the relations between the flexible board and the "reinforcing board" before drawing a hasty conclusion.

The flexible board that has drawn attention recently is as thin as paper compared with a hard copper-covered PWB in the past, is softer, hard to handle and more fragile.

The connector is soldered on the board like paper.

"Solder crack" is a possible defect caused by stress on the solder due to repeat attach and detach of the connector in the post-process and the market.

A "reinforcing board (0.2mm)" was put on the other side of the flexible board to cope with such a problem. Moreover, it is strongly attached with an adhesive so as never to peel off.

[Fig. 6-13] is an outline drawing showing that the flexible board with the reinforcing board was put in the printer.

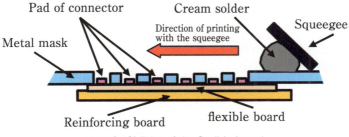

[Fig. 6-13] Print of the flexible board

The reinforcing board is placed at the bottom and the squeegee is on the top as shown in the outline drawing.

Between those are the flexible board and the metal mask (about 0.1mm-thick), which are soft. Imagine starting printing the cream solder that moves the squeegee from the right to the left.

At this time, the reinforcing board is slightly moved to the left when the printing pressure is set strongly.

In this case, because the metal mask doesn't move, the cream solder will be printed slightly on the right of the pad. (This is A in [Fig. 6-9])

When the squeegee moves from the left to the right, the state is

129

To the world of defect-free soldering

as shown in B in [Fig. 6-9].

Therefore, the cause of "Print misalignment" was the "**reinforcing board**". And the print misalignment has not occurred in chip parts without the reinforcing board.

"I see, but it's impossible to use the reinforcing board".

"It is possible if the present printing pressure is reduced. With the pressure, the print misalignment is totally eliminated".

"I see. This is exactly a mystery as the cause was the reinforcing board on the other side of the board.

However, I won't allow the same defect to occur in the new factory as long as I heard this story. I'm looking forward to going to the new factory".

《 Episode 3: The End 》

《 Quality improvement person's comment from Episode 3 》
A similar thing to this happens at your flexible board factory. In this case, pay attention to the printing pressure. However, each production site will need to decide its concrete numerical value.

6 Soldering mystery collections (Who is the criminal?)

(Episode 4) Does the characteristic of solder change suddenly!?

The man in a suit muttered while loosening his tie.

"Uh, I'm in trouble. Sales activities are really tough".

He is a familiar salesperson to me, and we often enjoy meals together.

"What happened to you?"

"The plant manager of our business partner suddenly complained about the solder we delivered the other day".

"Oh, really? What kind of complaint? "

He referred to our solder, "As soldering worsened, withdraw all the lots delivered and replace them with new ones".

"I see. That is unreasonable".

"Our company (Engineering section) analyzed the lots pointed out by the customer."

"What was the result?"

"They were the same as the old lots without any problem according to our engineering section. How can I explain to the customer's manager? I am in trouble."

"You're in a dilemma between the customer and your company, aren't you?"

"Anyway, the answer is on the site. There is no time. Let's go to the factory in question together now!!"

The customer's plant manager was unexpectedly gentle. Immediately, we were introduced to the person in charge of the site.

Here is the summary of our discussions with him.

1. The method of soldering is flow soldering (DIP soldering).
2. The problem is that there is a lot of defective soldering. In particular, solder doesn't adhere to a big part terminal. ('Soldering defect' of [Fig. 6-14])
3. Therefore, correction work (soldering iron work) is needed, and the production site is confused.

131

To the world of defect-free soldering

"I want to see the actual production site".

"Please come in".

The test was tried with one board, and it was soldered in the flow solder bath. The soldering was poor and defective as the person in charge said. In addition, a defective soldering also occurred in the second board.

[Fig. 6-14] solder lift failure

But, I felt doubt seriously.

《 Questions 》

1. The defects concentrated on the latter half of the board. The defect didn't occur in the front and the center in the direction of the board movement. (Refer to [Fig. 6-15])

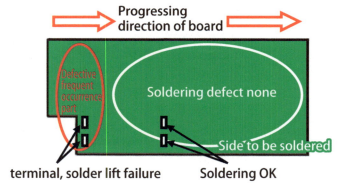

[Fig. 6-15] Defective generation part

2. It was soldering in the flow solder bath where the same "molten solder" was used. Therefore, it was unnatural that the defects concentrate on a certain part of the board.

I reached a conviction.

"Ladies and gentlemen, look at this".

6 Soldering mystery collections (Who is the criminal?)

"Defective soldering is seen only in these two terminals behind the board. These big parts have four terminals, and the front two terminals are soldered without trouble". (Refer to [Fig. 6-15])

"Oh! Surely".

"Don't you think that the key to the cause investigation lies here? Remove the cover of the flow solder bath and put the board!"

....

"Ah!!"

"Oh my God! Flux is not spread on the latter half of the board!" The latter half was soldered without flux.

Moreover, the composition of the flow solder bath is as shown in [Fig. 6-16] and the board passes on the flux sprayer, so flux is supposed to be spread on the entire board.

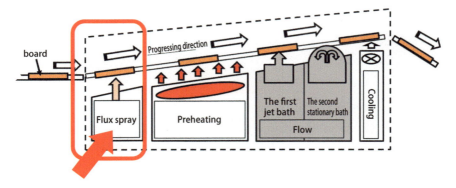

[Fig. 6-16] Composition of flow soldering equipment

The atomization of the flux is controlled by turning "on" and "off" with the sensor SW (switch), and the flux is sprayed onto the entire board.

The cause was easily found out.

Flux was applied from the sprayer to the board, but the position of the sensor was inappropriate and the latter half of the board was not atomized without flux as shown in [Fig. 6-17].

To the world of defect-free soldering

"However, what can we do with this?"

"Then, let's do this ……. Change the sensor position to atomize two more times".

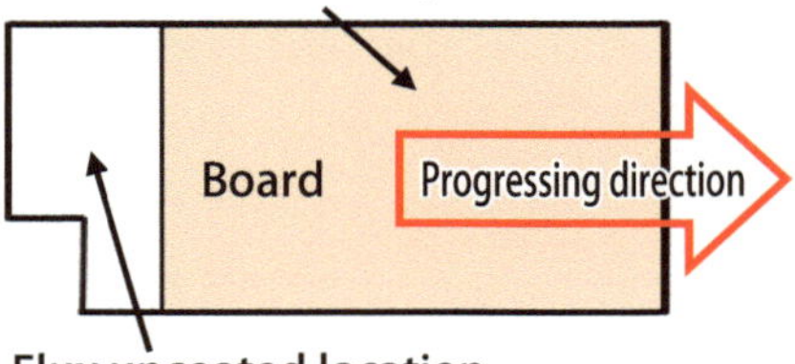

"Oh, that's nice!"

Of course, the test after the sensor position had been readjusted resulted in excellent soldering.

Flux had been spread

Board | Progressing direction

Flux uncoated location

[Fig. 6-17] Some flux uncoated

Thus, the awkward problem that had troubled him, a salesperson, was solved. The cause of the defect had nothing to do with the solder lot, and it was not necessary to replace it.

"Thank you so much, everything was solved without replacing the lot. Now, let's go for dinner together! I will pay the bill".

《 Episode 4: The End 》

《 Quality improvement person's comment from Episode 4 》
If the model changes, the size of the board is naturally different. So, the width of the conveyer chain is adjusted. However, the worker overlooks the sensor position. Ladies and gentlemen, because such a defect occurs when the model is switched, it is always necessary to adjust the sensor position.

6 Soldering mystery collections (Who is the criminal?)

(Episode 5) Mystery of terminal IC "No solder adheres"

One day in December

In front of the entrance door that had opened quietly, there was a familiar man who looked fine.

"Oh, Mr. plant manager, what's wrong with you?"

"I need your help again·········."

The voice of the plant manager seemed to be smaller than usual, and his proud beard looked sad.

I served tea to the plant manager who sat on the chair, and heard the circumstances from him.

According to him, the defects were as follows.

In the SMT production plant, 'No solder adheres' (defective rate: 1.2 %) of the SOP-IC terminal seemed to have occurred. (Refer to [Fig. 6-18])

Such a defect had not occurred in the past.

The IC is taped 12 PIN. According to him, the terminal of 'No solder adheres' is always a terminal on the far right.

On the site, after printing the cream solder, the terminal of SOP-IC doesn't contact a worker's hand in the process of being automatically mounted from the taping cassette with the mounter.

In addition, the print amount of the cream solder to the pad is proper, and there is no problem.

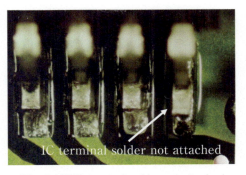

[Fig. 6-18] IC terminal solder not attached

"The cause is cannot be found because the defect does not occur frequently. A problem causing the defect was not found no matter how thoroughly the production site was checked while the mounter machine was operating. A small IC is picked up from the

To the world of defect-free soldering

taping with a nozzle and mounted on the predetermined point. Why cannot the cause be found in such a simple process?"

"Now, don't be disappointed, and investigate the cause".
I had proceeding to the factory with the got depressed plant manager.

"Are these actual defective articles?"
I carefully observed one defective article on the site with the plant manager.

In addition, I collected three defective goods and observed them in detail with a microscope to compare them with non-defective items.

....

"Mr. plant manager! Look at this for a moment".
I noticed strange symptoms of the three defective articles in comparison with the non-defective ones.

"Look at the three defective articles from above. Only the terminals of the three 'No solder adheres' items become a little longer than the other terminals". (Refer to [Fig. 6-19])

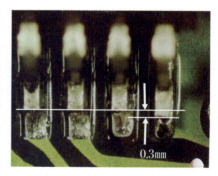

[Fig. 6-19] Size of IC terminal is uneven.

"Oh! I see. They seem to be a little longer as you say. Let's measure it with the microscope".

When the plant manager measured it with the microscope, the terminals of 'No solder adheres' were longer by 0.3mm.

"This is a big discovery!"
When I investigated this further, I noticed only these terminals

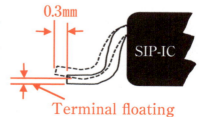

[Fig. 6-20] Smoothness is defective of the terminal.

136

6 Soldering mystery collections (Who is the criminal?)

were a little floating. (Refer to [Fig. 6-20])

"This is defective flatness of the IC terminals".

"Because the terminal is floating, 'No solder adheres' occurred".

I immediately checked the site, but the terminals in question did not seem to hit something or float at all.

"There is supposedly a problem with parts if there is nothing wrong with the site. Let's check with the IC maker."

The plant manager began to contact the IC maker immediately.

....

"According to the maker, there is not such a problem".

"I see. It is necessary for us to collect more defective articles to verify. Shall I check the taped ones containing an IC?"

"Are you sure?! All these? 2,000 pieces!"

"It is possible to grasp what is wrong if we see from above. It is possible to check them one by one with a ruler, although it is a little hard to see them due to a translucent taping cover".

When I began to check the 2,000 taped goods with the microscope, the plant manager also started checking.

For a while, we were looking for terminals with long ICs with microscopes.

"Oh, it's still barely 500!"

"It is still 1/4. When can we finish checking everything?"

"Oh, No! It is not necessary to see any longer. The conclusion is already drawn".

"What? We are still on the way".

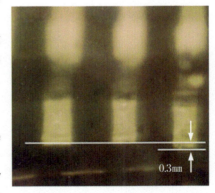

[Fig. 6-21] Defective symptom in state of taping.
It was discovered with state of cover of translucent.

137

To the world of defect-free soldering

"I found 4 pieces of ICs that were defective ([Fig. 6-21]) in the 500 pieces. There is nothing wrong with the SMT production site. Let's pass them to the IC maker for countermeasures without taking the translucent cover".

Then, the plant manager contacted the IC maker again. The taped defective goods were passed to the maker at a later date.

-- At a later date --

"Detailed explanations of the problem were given from the maker! After all, we were right!"

There were the following descriptions in the explanations from the maker in the hand of the factory manager who got excited.

《 IC maker's comments on cause and measures 》
The cause

This IC is sucked with the nozzle and packed to the tape automatically. It was packed with the tape being a little misaligned because the screw to fix the tape was loose.

At this time, it is thought that the IC terminals in question sucked with the nozzle hit the side of the tape, were deformed and floated.

The measures

The screw to fix the tape was tightened hard enough not to be loosened. Afterwards, there is no deformation or floating of the terminal.

"The cause was found and the measures were taken although it had taken time, and then there is no defect. It is really good".

Workers on the SMT site are happy again, increasing the production.

《 Episode 5: The End 》

Ladies and gentlemen, when a problem occurs, its cause will be found for sure if analyzed and verified one by one patiently and carefully. Your trouble disappears as long as the cause is found no matter how long it takes.

6   Soldering mystery collections (Who is the criminal?)

(Episode 6) Mystery of production plant late at night

Enjoy coffee after the meal. There was unusually no client, and it was quiet today.

By the way, a mysterious event occurred in the production plant late at night before.

"It is a good opportunity to talk about this to you."

"It is a story of a flow soldering site at a certain factory."

The factory operated round the clock with 3 shifts.

There were many defects with boards produced during the shift from the night before when the 8 AM shift started.

Somehow, the defects occurred late at night.

The morning shift workers of the following day were confused and became very busy with adjustment and repair of the defective boards produced during the night shift, resulting in worsened manufacturing efficiency.

Why does such a problem happen in the same production site only late at night, although the workers are different? Why late at night?

It is mysterious.

Days of confusion and uneasiness continued on the site.

I was called to review the work of the night shift in such a situation.

Just in case, ghost is not my expertise.

Well, it was my second day on the production of the same boards both day and night without being

haunted by ghost late at night.

The trouble with the boards was insufficient soldering as usual.

When I opened the flow solder bath and searched for the inside, there were few flux liquids in the container.

There was damage in the tube where the flux liquid was injected,

139

To the world of defect-free soldering

and the liquid kept leaking from the tube.

The amount of flux was insufficient late at night even if it was thought that there was a sufficient amount of flux liquid.

Soldering was carried out without flux and, naturally, there was defective soldering.

Immediately, the tube was replaced with a new one, and flux was sufficiently applied.

It is necessary to think about countermeasures for the future, although the problem was solved as above. The equipment has to be checked during daytime because the number of workers decreases late at night.

There is no ghost in the production plant late at night.

《 Episode 6: The End 》

It is important to confirm there is an amount of liquid in the container enough to use for the production beforehand. In addition, it is recommendable for the on-site QC patrol (ISO) member to add this item to the check sheet and check it regularly. For you not to be annoyed by ghost.

6 Soldering mystery collections (Who is the criminal?)

(Episode 7) Suddenly, 'board swelling' happens

"Hello". One day, a middle-aged man opened the door and came into my room.

"I plan to retire from the company in which I have worked for years this time, and to return to my hometown. I wanted to meet you again by all means".

He began to reminisce about old days nostalgically while drinking coffee.

"When I joined the company, most boards were single-sided PWBs. After a while, they were replaced with double-sided PWBs. The double-sided ones were thought to become mainstream, but final products have been miniaturized and been of high density, and boards themselves have become multi-layered" (Refer to [Fig. 6-22]: 7-layer board)

[Fig. 6-22] Section of multilayer board

"Various kinds of equipment were introduced to the production site for the multi-layered PWBs, replacing conventional production facilities and confusing the site".

"I certainly understand. Because time has dramatically changed, you supposedly had a hard time also on the site".

"By the way, 'board swelling' suddenly occurred in the multi-layered PWBs one day when we were struggling as such every day". (Refer to [Fig. 6-23])

"Because 'board swelling' was the first case for us, my site was very confused".

The man kept talking further.

"Even if the temperature profile of the reflow furnace

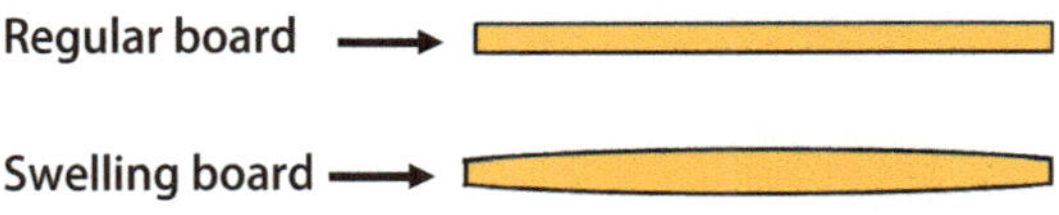

Regular board ⟶ ▭

Swelling board ⟶ ▭

[Fig. 6-23] Swelling symptom of board

141

To the world of defect-free soldering

where the temperature was added was checked, the peak temperature was 245℃ as usual without problem".

"I requested the board maker to take measures by concluding the board itself was defective".

"The maker investigated various possible causes, but there was nothing different from conventional boards after all, and its answer was that it was not a problem of the maker".

"The problem still continued after that. We checked our temperature profile and the production process, but no problem was found. Therefore, I doubted the board maker even more".

"At that time, a young staff in the Manufacturing Department said the following".

"Is the temperature profile of the reflow furnace really correct?"

"I thought at first, 'We have carefully checked the temperature profile! What such a youngster understands! It is clear that the cause is in the board maker!' But the young staff began to check a board sample with a thermocouple whose temperature profile was measured after returning to the site with the reflow furnace".

"When he showed the sample to me, (the sample : board sample to perform temperature measurement) it was not possible for me to say anything." ([Fig. 6-24] **A**)

[Fig. 6-24] Adhesive too much   [Fig. 6-25] Proper quantity of adhesive

"I was deeply sorry for the criticism to him later at that time".

The result of the young person's investigation was as follows.

"When I saw the temperature profile ([Fig. 6-26] **a**) at the

6 Soldering mystery collections (Who is the criminal?)

beginning, I noticed something strange".

"The curve of the peak of the temperature (about 245℃) was greatly different compared with the conventional temperature profile".

"It is a problem that the temperature curve in the peak is round (R is large)".

"There is too large an amount of heat stiffening adhesive for attaching the thermocouple shown in ([Fig. 6-24] **A**) to the board".

"When a lot of adhesive is used, a low temperature is displayed on the profile in the reflow furnace due to poor heat conduction".

"If so, the internal temperature of the actual reflow furnace is supposed to become too high".

A new sample ([Fig. 6-25] **B**) was urgently produced with his guidance, and we took the temperature profile once again.

We were surprised to see the profile. Surprisingly, the peak temperature had gone up to 300℃!

As a result, the cause of 'swelling' of multi-layered PWBs .lied in the board sample for measuring the temperature profile shown in ([Fig. 6-24] **A**). Because the sample was bad, the temperature of the reflow furnace has been raised up to 300℃. Still, the profile of the temperature recording form displayed the peak temperature as 245℃, so we misunderstood that we were not wrong. (Refer to [Fig. 6-26] for comparing profiles of the board having the problem and the non-defective board)

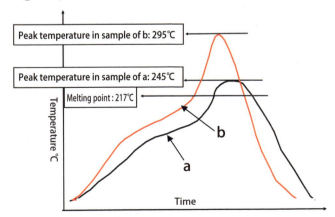

[Fig. 6-26] Temperature profile comparisons of boards

To the world of defect-free soldering

Afterwards, we learnt how to make the thermocouple and the board for temperature survey (below) from him.

"In the temperature profile, the correct measuring method is to measure the contact point of the thermocouple (two main lines) and fix the point with an adhesive temporarily".

"Unless making a sample (board) for temperature profile measurement as close to as mass-produced boards, it is impossible to make an accurate temperature profile".

"Because the temperature profile also means "written confirmation of the temperature", it is very important".

"The cause of this problem was a sample for temperature profile measurement, and the problem was successfully solved. 'Board swelling' has disappeared since the temperature setting was changed with a sample made with a smaller amount of adhesive".

"It is natural that the board got angry to be 'swelling' if the peak temperature was raised to 300℃".

However, the youth left the factory for some reason and is currently missing.

《 Episode 7: The End 》

The temperature profile is posted in the reflow furnace at each factory. The currently produced board is posted as "confirmation" that they are produced under this temperature condition.

Moreover, the temperature profile frequently draws attention in a quality meeting in terms of defects and temperature conditions at the time when they occurred. Even such an important temperature profile is meaningless if the base samples are far different from the mass production stage and untrustworthy. When the sample boards are made, it is necessary to make them under conditions as close to as those of the mass production.

6 Soldering mystery collections (Who is the criminal?)

(Episode 8) The display of the timer disappeared suddenly three years after buying!?

"Ladies and gentlemen, the timer is very convenient. Do you have it? What? You don't ···."

"I'm a night person and cannot wake up early in the morning".
"Therefore, I bought the digital display timer made by company A. The timer can work very well as an alarm clock".
"I set the timer at six o'clock every morning when the news program started, and it was convenient".
"I woke up for sure and was never late for work".

"However, the display disappeared suddenly one morning about three years after I bought it".
"It is ridiculous to buy another timer".
"As I wanted to investigate the cause of the breakdown, I checked the board inside the timer for the first time".

"I was really surprised!"
"The pattern that was supposed to be there disappeared!"
"Do you see? Here!"

"Do you think that such a thing can happen on the pattern side ([Fig. 6-27]) of the board?"

"I have soldered on production sites for 40 years, but it is the first time for me to encounter such a case as a copper foil pattern of the board completely disappeared in a commercially-available product".

"This is truly a mystery".

Purchased timer

What! Pattern disappeared (melted)

[Fig. 6-27] the pattern was lost

145

To the world of defect-free soldering

"However, as a result of various investigations, we reached a surprising conclusion".

"What! The electrolytic capacitor caused the problem".

"The liquid of the electrolytic capacitor that was attached to the part opposite to the soldered side had leaked".

"In three years of time, the liquid leaked and melted the pattern little by little".

"Once the cut pattern was connected with the electric lead, the timer began displaying time and the problem was solved".

"Still now, this timer is working without problem".

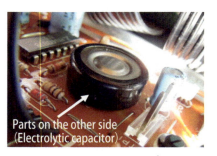

[Fig. 6-28] Photo of criminal's face

《 Episode 8: The End 》

"Ladies and gentlemen, we will recommend you to check the inside of a broken electric product. Something unexpected may be discovered".

6   Soldering mystery collections (Who is the criminal?)

(Episode 9) Suddenly, a pattern is lost.

"Wow! I was really surprised!"
"What? What happened?"
"Recently, the pad disappeared suddenly, and I wondered why".

He is young with brown hair.

He is more used to the work and is reportedly working at the flow soldering site now. He seems to make efforts in his own way.

According to him, most parts of the pad disappeared as shown in [Fig. 6-29] on his site during production of certain boards.

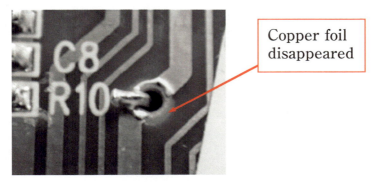

[Fig. 6-29] Copper foil disappeared

"I do not understand why such a thing happened".

I saw the photograph showing this problem, and discovered a past similar photograph. I drew the following conclusions.

It usually takes 4-5 seconds for the board to pass the flow solder bath. However, there is a trouble that the board stops suddenly on the flow solder bath.

The copper pad portion begins to melt more and more to the flow solder bath, and the pad gets smaller if the board stops long.

The copper foil of the board diffuses to the molten solder bath.
This is called, 'Copper foil disappearance'.

The cause that the board stops suddenly on the way is long terminals of the insertion parts or poor adjustment of the width of the

To the world of defect-free soldering

chain.

In that case, the board remains stopping until the following board's coming and pushing it out.

If the following board doesn't come, it is the worst.

Meanwhile, the copper pad begins to melt more and more to the solder bath.

It is very important for the person in charge to regularly check the inside of the flow solder bath from time to time.

《 Episode 9: The End 》

Ladies and gentlemen, bear it in mind that it is necessary to sometimes open the door of the flow solder bath to check the direction of a board.

6 Soldering mystery collections (Who is the criminal?)

(Episode 10) Defective goods are full of treasures! (Important know-how hidden there)

Today's client is a longtime friend of mine.

Hmm? What kind of person is he?

If you meet him, you understand how he is like at once. He will come soon.

Apparently, he seems to have arrived.

A big black car stopped in front of my house. Well, I go to see him.

"I haven't seen you for a long time".

"I just want you to know my trouble".

"I want to hear your opinion on the trouble in our factory".

This man is a longtime friend of mine and is a president of a company having a big production plant.

"Well, just calm down. Let's talk after getting in my house".

He started talking about the problem while drinking coffee I had served.

"The target of my production site is zero defect! A defect not only requires excessive costs and time, but also causes loss of trust of the customer as a matter of course. A defect is simply all pain, no gain!"

"However, the number of defects is decreasing in our company, but they still occur".

"The defect as shown in [Fig. 6-30] occurred in the process of the customer recently, causing its complaint".

"This board (PWB) was a new product but its design changed with a partial change of parts and additional work. So, we changed chip parts and added an electrolytic capacitor by using a soldering iron. Therefore, gloss of the flux residue partially remained on the board".

"This board was confirmed as a defective product by the customer although our company shipped it as a non-defective product. The lead of the electrolytic capacitor being soldered to the IC terminal

To the world of defect-free soldering

came off".

"Hmm, please let me see the photograph".

I looked at it carefully and pointed out the following things after a while.

[Fig. 6-30] Example of soldering defect

《 3 points found while looking at the photograph 》

1. The flux residue has adhered to the board.
   (Since there was a design modification, there is a trace of having attached the mini diode, the resistor, and the big electrolytic capacitor with soldering iron work)
2. There is a big difference in gloss of wire solder used in the mini diode terminal and in the lead of the electrolytic capacitor. There is no solder gloss of the lead of the electrolytic capacitor. (The former had peculiar gloss to solder without problem, but gloss was not seen in latter. Since the worker spent too much time for the latter, the flux already evaporated, and there was no gloss. The solder became fragile in violation of the 《 4

6 Soldering mystery collections (Who is the criminal?)

soldering elements 》)

3. Although it looks as though it seems to be attached with no problem in the photograph, actually the soldering part between IC lead and electrolytic capacitor lead was peeled off in customer's process. No solder fillet is formed on the lead of the peeled electrolytic capacitor. (Refer to the red circle in [Fig. 6-30]) The absence of solder fillet means that there is no trace of trying to raise the temperature without putting a soldering iron tip on the lead part. Therefore, due to insufficient temperature, alloy layer could not be formed. Ultimately, what was stated in **2** and **3** is a work that is contrary to "four elements of soldering", that is, soldering without flux and soldering without adding temperature to the metal, so that it becomes defective in the customer's process. (Refer to [Fig. 6-31])

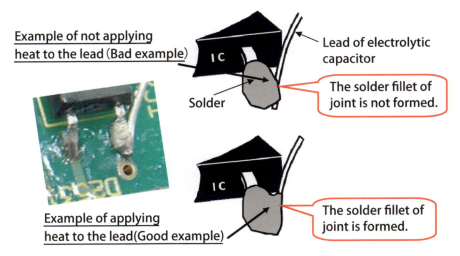

[Fig. 6-31] In the above figure, heat was not added to the lead and in the below figure, heat was added to it.

The main purpose of a soldering iron is not to melt the wire solder, but to raise the temperature of the base metal. (Refer to pages 27 and 28 of the text) However, it was only to insert a lead to molten

To the world of defect-free soldering

solder without applying the tip of the soldering iron to the lead. In this work, the fundamentals of soldering iron work were ignored.

"It is a photograph that shows work being done in contradiction to the basic soldering principles. Don't you think so?"

"Yes, you're right".

"This is a case where 'two elements' of the 'Four elements of soldering' were missing".

"They are 'Hot (temperature) was not added to the lead' and 'Flux evaporated and already did not exist'".

"Yes. unless paying attention to the temperature relation of the 'Florida pattern' (refer to page 61 of the text) in reflow soldering and flow soldering, the soldering iron work causes the defect as shown in the photograph".

"The important is **'temperature'**. It is not easy to pay attention to temperature as it is not visible. This is the root cause".

I think that customer complaints and defects occurring on the production site are full of treasures that are very meaningful for the relevant persons. The treasures are demanding us to investigate the cause of the defect and take measures soon. They shine like diamonds.

《 Episode 10: The End 》

Ladies and gentlemen, think about it carefully. For example, a defective rate of 1% means one defect among 100 items produced. There was a difference of a "cause of something" between 99 non-defectives and 1 defective. If the "cause of something" is investigated and "countermeasures" are taken, the defect rate is supposed to become 0%. Moreover, the defective one surely has a "cause" of being different from the 99.

Defective goods having the "cause" shine brightly as diamonds. Don't you feel that they demand us to find the "cause" soon?

I think that it is useful to master 'Basics of the soldering iron work', 'Skill training', and 'Basic knowledge of soldering' to avoid a customer complaint.

(Exactly, this defect is a treasure teaching us various things)

6 Soldering mystery collections (Who is the criminal?)

(Episode 11) Automatic driving, is it okay?

The car society is now trying to make major changes and advances.

Historically, the era of using gasoline engines has long continued, and recently, many "electric cars" have come to the market to prevent global warming.

The electric cars are also becoming very con-venient with various functions, and there are also cars with great functions such as automatic brakes and automatic steering wheels.

And now, looking at the world in an aging so-ciety, it is predicted that electric cars with the function to drive when elderly people feel anxious about their driving technology will be widely marketed.

I think it would be great if electric cars prevent the elderly people from making mistakes and eliminate the disastrous traffic accidents that have become a major social problem.

However, it is regrettable that the accident will not disappear even with an electric car (with an automatic brake function).

In Japan, with electric cars there were about 340 accidents in 2018.

When an electric car with an automatic braking function causes a traffic accident, it becomes a problem whether the driver is responsible or the car manufacturer.

However, I have the following concerns.

It is natural for electric cars, but many electric signals are used to control convenient func-tions, and for that reason, many printed circuit boards (chip parts, transistor parts, IC parts, etc.) are used in cars.

All of these electrical components are auto-matically soldered.

The car has an environment with a temperature difference of about 100°C between midsummer and midwinter, and is also subjected to vibra-tion when traveling on a road (rough road).

Do not forget that you are riding an electric car with an automatic driving function and an automatic braking function, and you are driv-ing

153

To the world of defect-free soldering

with temperature test and vibration test even if you are driving safely.

The weak part of many mounting boards under the temperature test and the vibration test is the soldered part.

Even if the manufacturer does automatic sol-dering, it is highly possible to perform reliable soldering according to the basics of soldering. It will be important.

If there is no problem in 1 or 2 years with the fake soldered boards, the electric signal can't be controlled after 3 or 4 years, and a traffic ac-cident will occur. I would like you to be very careful as a manufacturer.

《 Episode 11: The End 》

《 The author's self-introduction 》

I graduated from the Department of Electronic Engineering, Osaka Institute of Technology in 1970, and joined ALPS Electric CO., LTD.

I had worked at ALPS Electric CO., LTD. for 30 years, afterwards for about 10 years in a Chinese company which was said to be "Factory of the world".

I belonged to the engineering department, the production engineering department, the quality control department, the production department by a company work for 40 years, and I engaged in duties related to "soldering" mainly.

When I belonged to the engineering department, I always thought about how the soldering without the problem should do a pattern design.

To 'improve soldering quality and take measures against defects' on a production site, I have analyzed a thorough cause of a defect one by one and taken measures against it. (The production site is for SMT reflow soldering, flow soldering of insertion parts, and work with a soldering iron)

Meanwhile, I acquired four qualifications of The Japan Welding Engineering Society (for operator, for quality inspection personnel, for instructor, and for engineer). Currently, I have only the one for instructor as a result of updates.

While I was handling a process of: (occurrence of defect) → (cause analysis) → (measures) → (confirmation of recurrence prevention) on production sites many times, 'soldering' that had been a spot at first gradually became a line, such lines gathered to make a plane, and my

To the world of defect-free soldering

hobby had become 'soldering' at last.

And, I still get excited for 'O soldering defect' whenever I stand beside each machine on the SMT production site, and cannot stop my heart and mind from burning.

This time, I decided to summarize examples of measures against defects that occurred on the production sites in this booklet as 'Knowhow of Solder without Defect' after having consulted with persons concerned.

This is a collection of measures as a result of my thorough investigations of defective goods and analyses of the causes.

From my 40-year experience in soldering, I strongly feel it is very important to thoroughly investigate defective goods and analyze the causes, and I am working on the site now.

I am grateful if this booklet could help you even a little on your production sites.

Lastly, I would like to sincerely express my gratitude for the cooperation of parties concerned in ALPS ELECTRIC CO., LTD., SENJU METAL INDUSTRY CO., LTD., and HAKKO CORPORATION, and colleagues in China for being involved in measures against defects.

This booklet was finally completed thanks to everyone concerned. Thank you very much.

To the world of defect-free soldering
Measures against defects that can be made at production sites

Copyright ©2019 by Minoru Kishindo
All rights reserved.
No part of this publication may be reproduced, stored in a retrieval system, or transmitted in any form or by any means, electronic, mechanical, photocopying, recording, or otherwise, without the prior permission of the publisher.

Published in Japan.   ISBN978-4-86584-409-2

For information contact : BookWay
ONO KOUSOKU INSATSU CO.,LTD.
62, HIRANO-MACHI, HIMEJI-CITY, HYOGO 670-0933 JAPAN
(Phone) 079-222-5372 (Fax) 079-244-1482